THE DEFINITIVE GUIDE TO SUPPLY MANAGEMENT AND PROCUREMENT

THE DEFINITIVE GUIDE TO SUPPLY MANAGEMENT AND PROCUREMENT

PRINCIPLES AND STRATEGIES FOR ESTABLISHING EFFICIENT, EFFECTIVE, AND SUSTAINABLE SUPPLY MANAGEMENT OPERATIONS

Council of Supply Chain Management Professionals

Wendy Tate

Vice President, Publisher: Tim Moore
Associate Publisher and Director of Marketing: Amy Neidlinger
Executive Editor: Jeanne Glasser Levine
Consulting Editor: Chad Autry
Operations Specialist: Jodi Kemper
Cover Designer: Chuti Prasertsith
Managing Editor: Kristy Hart
Project Editor: Deadline Driven Publishing
Copy Editor: Apostrophe Editing Services
Proofreader: Apostrophe Editing Services
Indexer: Angie Martin
Compositor: Bronkella Publishing
Manufacturing Buyer: Dan Uhrig

Published by Pearson Education
Upper Saddle River, New Jersey 07458

For information about buying this title in bulk quantities, or for special sales opportunities (which may include electronic versions; custom cover designs; and content particular to your business, training goals, marketing focus, or branding interests), please contact our corporate sales department at corpsales@pearsoned.com or (800) 382-3419.

For government sales inquiries, please contact governmentsales@pearsoned.com.

For questions about sales outside the U.S., please contact international@pearsoned.com.

Printed in the United States of America

First Printing December 2013

ISBN-10: 0-13-344901-7
ISBN-13: 978-0-13-344901-3

Pearson Education LTD.
Pearson Education Australia PTY, Limited.
Pearson Education Singapore, Pte. Ltd.
Pearson Education Asia, Ltd.
Pearson Education Canada, Ltd.
Pearson Educación de Mexico, S.A. de C.V.
Pearson Education—Japan
Pearson Education Malaysia, Pte. Ltd.

Library of Congress Control Number: 2013952809

I would like to dedicate this book to my two daughters, who are always supportive of the things that I do. They helped me think through some of the issues associated with purchasing and supply chain management that I would never have considered. Whitney and Tayler Tate make my job more interesting. I'd also like to acknowledge the support of my immediate family in generating ideas for this book: Jeff Crow, Brent Crow, Sharon Crow, Julie Barry, Brooke Arnone, Dennis Crow, and Ruth Sykes.

I'd also like to acknowledge the hard work of two MBA graduate assistants. Anthony Mubarak and Prannay Ved helped to make this book possible with their fast responses and support of the project.

Finally, I'd like to thank the Council of Supply Chain Management professionals and Pearson Publishing for their support of this project and the supply chain management field in general.

CONTENTS

ABOUT THE AUTHOR

Wendy L. Tate, Ph.D. (Arizona State University, 2006) is an Associate Professor of Supply Chain Management in the University of Tennessee Department of Marketing and Supply Chain Management. She joined the faculty in August of 2006. She teaches both undergraduate and MBA students strategic sourcing and manufacturing and service operations. She has an interest in the cost impacts of business decisions across the supply chain. She is active on the UT campus and serves on many committees. She served for five years as the faculty advisor of the Council of Supply Chain Management Professionals (CSCMP), an active student-run professional organization for supply chain managers. She has published in both academic and practitioner top-rated journals in supply chain management. She enjoys research and takes a special interest in translating academic work into classroom learning activities.

Her research focuses on two primary business problems. The first is in the area of services purchasing, including outsourcing and offshoring. The second is on environmental business practices and trying to understand how these initiatives can be diffused across a supply chain and a supply network. She presents at many different venues including both academic- and practitioner-oriented conferences.

Wendy has two girls, Whitney and Tayler, who are both students at the University of Tennessee in the business college. Wendy enjoys the time she spends with them. She also enjoys gardening (she is currently working on a Landscaping Certification), traveling, hiking, and reading. She frequently attends cultural and sporting events. She also interacts with the literacy program in Knoxville and enjoys teaching students at all levels how to read and especially how to understand written problems so that they can potentially obtain college or technical degrees.

1

THE ESSENTIAL CONCEPTS OF PURCHASING AND SUPPLY MANAGEMENT

The chapter discusses the roles and objectives of procurement in an organization and in supply chain operations, describes the different types of spending and buying methods, and defines key terminology. Supply management is an important business function that makes a significant contribution to both the top and the bottom line of the organization. The supply chain starts with finding, selecting, and managing effective and efficient suppliers of materials, equipment, and services. The capabilities of these suppliers are an important asset to the organization in terms of innovation and customer value.

Learning Objectives

After completing this chapter, you should be able to:

- Understand the differences and appreciate the evolution from tactical purchasing to strategic supply management.
- Understand the purpose and goals of the supply management organization and its contribution to profitability and the bottom line.
- Identify the relationship between the purchasing function and other functional areas.
- Identify the activities that are part of supply management.
- Understand the different categories of purchasing spend.

Introduction and History of Purchasing

Purchasing is one of the basic processes common to all organizations. It is the process of acquiring goods, services, and equipment from another organization in a legal and ethical manner.[1] Purchasing was initially a tactical contributor to the organization, focusing on transactional relationships and low price (Table 1-1). However, over time the role of the purchaser, and the purchasing department, has changed significantly and the function has become strategic to organizational competitiveness.

Table 1-1 History of Purchasing[2]

Period	Status
Late 1890s	Purchasing rarely used as a different department except in the railroad.
Early 1900s	Purchasing considered clerical work.
World War I and II	Purchasing function increased in importance due to the importance of obtaining raw materials, services, and supplies to keep the mines and factories running.
1950s and 1960s	Continued to gain stature, processes more refined, and more trained professionals. Still considered order placing clerical in a staff-support position.
Late 1960s–early 1970s	Integrated materials systems introduced, materials became part of strategic planning, and importance of department increased.
1970s	Oil embargo and shortages of basic raw materials turned the focus of the business world to purchasing.
1980s	Advent of just-in-time with an emphasis on inventory control and supplier quality; quantity, timing, and dependability made purchasing a cornerstone of competitive advantage.
Early 1990s	Value proposition of purchasing continued to increase; cost-savings became the buzzword.
Late 1990s	Purchasing evolved into strategic sourcing, contracts were more long term, and supplier relationship building and supplier relationship management started.
2000s	Purchasing shifted its myopic focus on cost to much broader terms. Some of the widely used developments: spend analysis, low-cost country sourcing, procurement technology evolved (ERP, e-sourcing), procurement outsourcing evolved (P2P), total cost of ownership, data mining and benchmarking, and lean purchasing.

Globalization has forced companies to improve their internal processes, such as supply management, to remain successful. The level of competition in the marketplace expanded to include both domestic and international markets. Purchasers no longer discuss "lowest price" but share information, collaborate, and talk to their suppliers about total costs, life-cycle costs, and cost reductions. This requires a focus on process improvements instead of short-term relationships and price reductions.

The primary goal of the purchasing organization is to purchase the right item or service, in the right quantity, at the right price, and at the right time. An abundance of competitors and seasoned customers demand higher quality, faster delivery, and products and services customized to their needs, at the lowest total cost. These demands are made at an even greater speed because of the influx of technology and social media into business-to-business applications. Information and data flow between supply chain members are increasing, making it challenging for organizations to continuously adapt to the ever-changing needs of the customer.

Getting products to customers at the right time, place, cost, and quality constitutes an entirely new type of challenge. Technology and an improved logistics network have opened up a world of opportunity to better enable competition through an expanded, globally oriented network of suppliers. The availability of low-cost labor and other alternatives in emerging countries has led to unprecedented shifts toward outsourcing and offshoring (discussed further in Chapter 5). China has become a major world competitor, introducing even more challenges for United States organizations in both the manufacturing and services sectors. The services sector now accounts for approximately 70 percent of the Gross Domestic Product (GDP), introducing more opportunity for effective supply management involvement in this sector[3] (Table 1-2). In the manufacturing sector, the vast majority of materials are purchased from sources outside the firm. Because of this, the supply management function has grown in importance and in complexity.

Table 1-2 U.S. GDP and Private Services-Producing Industries[4]

Millions of Dollars	2008	2009	2010	2011	2012
U.S. Gross Domestic Product	$14,291,543.00	$13,973,681.00	$14,498,922.00	$15,074,667.00	$15,684,764.00
Private Service-Producing Industries	$ 9,715,905.00	$ 9,609,550.00	$9,968,918.00	$10,357,395.00	$10,778,324.00
	68.0%	68.8%	68.8%	68.7%	68.7%

Case in Point—Finding the Right Suppliers

Purchasers must locate the right suppliers for their products and services. The issue is not so much about locating suppliers, with so many suppliers located in a variety of regions around the world, but about ensuring that they can meet your business requirements. Over time, these business requirements have changed significantly and purchasers have had to adapt their selection process to include criteria that often are difficult to truly assess. For example, recently apparel manufacturers in Bangladesh were lacking fire escape doors (a safety feature), leading to the deaths of many employees. Companies have entered into contracts with suppliers in some regions that tried to "bribe" officials into paying additional money to manufacture their goods. There are many instances in which the threat of Intellectual Property (IP) theft caused companies to eliminate certain regions of the world from supplier consideration and manufacturing location consideration. There are many other highly publicized examples in which the wrong supplier was selected because the quality of the supplier was difficult to assess. Purchasers have to know what to look for and have to ask suppliers some potentially difficult questions.

All these changes and challenges have helped propel supply management to the forefront of strategic decision making. The importance of appropriate management of suppliers that provide materials and services has become a key consideration for executives. There are many instances seen daily in which companies have received negative publicity due to the actions of suppliers, the location of suppliers, or the performance of suppliers. Competition is now between supply chains.[5] The companies that configure the best supply chains, with a highly performing supply base, will be the market winners and gain competitive advantage.

Why Is Purchasing Important?

Historically, purchasing has played a key role in "getting the lowest possible price." This was often at the expense of a positive relationship with the supplier and usually a trade-off with quality. However, over time, the role of purchasing evolved into a cost-saving function in which supplier relationships and contracts were developed with cost-savings in mind. These cost-savings often came through process improvements, product improvements, or supplier development efforts.

Today, purchasing is recognized as having an overwhelming impact on the bottom line of the organization. It has a direct impact on the two forces that drive the bottom line: sales and costs. Purchasing is becoming a core competency of the firm, finding and developing suppliers and bringing in expertise that is highly valued by the organization. Purchasing is generally responsible for spending more than 50 percent of all the revenues the firm receives as income from sales. More money is often spent for purchases of materials and services than for any other expense, and the spend in services is rapidly increasing.[6] Often, the cost of materials is 2.5 times the value of all labor and payroll costs and nearly 1.5 times the cost of labor plus all other expenses of running the business.[7] In the area of services, millions of dollars are spent on marketing and advertising, legal, information technology, logistics, temporary labor, and other categories. Although the involvement of purchasing in the services area is different than in a typical purchase of materials, there is significant opportunity for most organizations to save money by involving purchasing in this area of spend.

Figure 1-1 shows how supply management can drive sales up and costs down. The impact on net income and return on investment (ROI) have a major influence on shareholder value. The cost impacts are easily understood because cost reduction is typically considered a "purchasing job." Purchasing works with internal customers to help improve processes and drive down costs. Purchasing also works with suppliers to improve processes, look at alternative materials, and look at different locations or transportation modes. Focusing on cost improvement is a core competency of purchasing professionals.

However, there are also many opportunities to help drive up market share. For example, strong relationships with the right suppliers might allow for early supplier involvement in new product development. Therefore, the supplier is prepared for the actual launch and can also contribute and make changes if appropriate to facilitate an easier and less costly production launch. In 1998, for example, suppliers were involved in the product development process and actually helped by providing inputs into the design of the Honda Accord. These inputs were both material in nature and process-oriented. This early involvement in the product development stages helped to save more than 20 percent of the cost of producing the car.[8] In the casting industry, it was found that early involvement of suppliers in product development saved time and cost, and improved the quality of the parts.[9] The request for quote process (RFQ) is reduced significantly in this industry because suppliers are more aware of what is required, long before it is needed.

As you will begin to understand from the information presented in this book, the supply base is a source of innovative opportunities and the supply manager is trained to be aware of these supply market opportunities. Having the appropriate supply base and relationships with the suppliers is like having thousands of additional people thinking of the next great idea or innovation. There is a famous and highly publicized quote about suppliers and the supply base by Dave Nelson, who was an award-winning purchaser who worked at Honda, John Deere, and Delphi—he is considered a "guru" of supply chain.[10] Nelson said, "If you develop the right relationship with your supply base, you can have 10,000 additional brains thinking about ways to improve your product and generate cost-savings."[11] There is a lot of power resting in the hands of supply managers, if they can harness the strength and the capabilities of the supply base.

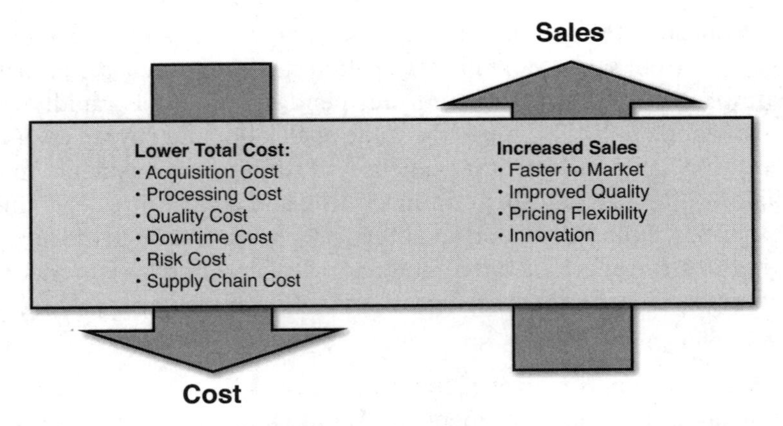

Figure 1-1 Supply Management's impact on both top and bottom line

How Can the Appropriate Relationships with Suppliers Create Value?

Purchasing can use its relationship with suppliers in many ways to help improve organizational value. This function has the capability to develop the appropriate relationships with suppliers that then become a competitive advantage for the firm.

Building Relationships and Driving Innovation

As purchasing continues to evolve as an important contributor to organizational success, the traditional approach of driving a hard bargain for price reductions has changed into a more relational approach with suppliers. This means that purchasing develops strong relationships with suppliers to jointly pull costs out of the product or service. This may be in the form of material changes to the product, process changes at the supplier's facility, raw material changes, or potentially process-oriented changes in transport.

It is this joint effort between the buying firm and supplying firm that maximizes the benefit. The expectations for suppliers continue to evolve as well. For example, suppliers are expected to, and generally are measured on their ability to, contribute innovative ideas, such as process improvement or cost reduction, which continually add value to the firm's products and services. Buyers are constantly searching to attract, retain, and maintain the best suppliers from around the globe.

Case in Point—Suppliers Driving Innovation

A supplier to a furniture manufacturer saw an opportunity to reduce waste and save on material costs. The change involved saw blades at the cutting station. The supplier suggested that the manufacturer reduce the width of the saw blade by 1/16". This decreased the amount of waste (sawdust) that was created from each cut. There were a number of other calculated savings from this small change given that the amount of raw lumber that was cut annually was significant. The same supplier also developed some alternative uses and sources for the waste material. Some of the material was sent to a manufacturer of particle board by developing a system that transported it directly from the cutting line into a rail car. Other smaller pieces of wood were basically glued back together and sold as "fingerjoint;" at one point, an entire line of furniture was produced using this fingerjointed (or waste) lumber.

Improved Quality and Reputation

Organizations are focusing on core competencies and their own areas of specialization. Because of this, outsourcing is increasing on both materials and services to find suppliers with core competencies that complement the firm. The relationships among purchasing, suppliers, and quality is becoming even more important than it was previously. These important relationships cross multiple tiers of suppliers, and lapses in managing supplier quality can tarnish a firm's reputation. The appropriate level of supplier quality can help to improve market share and increase sales.

Case in Point—Quality and Reputation

There were several different recalls involving pet food in 2013.[12] The first was a recall by a supplier of poultry feed. The danger was that the poultry feed had too high of levels of calcium and phosphorus, requiring the feed manufacturer to voluntarily recall its products.[13] The chickens and turkeys that consumed this feed were poisoned, and many died. The poultry farmers were advised on how to manage the problem and what to look for, but in some cases the expense and loss of inventory was challenging.

Another recall in August was issued for Iams and Eukanuba brand dry dog and cat food.[14] The food that was recalled was potentially harmful to both pets that ate the food and humans that handled the food. Pet owners were concerned, especially after the recall scare a few years ago that killed a number of their beloved animals. The food was possibly contaminated with salmonella.

Reducing Time to Market

Marketing research has shown that the first firm to introduce a successful new product will hold more of the market share after competition enters the picture. So, the first to market wins the majority of the market share.[15] Purchasing plays a key role in reducing the time to market by cross-functionally participating on product and service development teams. In many cases, carefully selected suppliers, as previously mentioned, are part of this development team.

Purchasing acts as the liaison between suppliers and engineers, trying to minimize engineering changes after introduction, and purchasers can also help to improve product and process designs by potentially suggesting standardized components or alternative materials, for example. An organization's ability to introduce new products can be constrained by a supplier's ability to meet demand, so suppliers have to be ready, willing, and able to ramp up production of new products. The development and management of competent, responsive suppliers are critical to get products to market quickly and also to respond to increasing or fluctuating demand.

Case in Point—Early Supplier Involvement[16]

Companies that involve suppliers in the early stages of the product life cycle (for example, during design inception) can reduce product development cost by 18 percent and improve their time-to-market cycle by 10 percent to 20 percent.[17] Best-in-class companies understand that suppliers play key roles in enabling them to successfully implement their products. There have been estimates that 80 percent of a product's cost is fixed by the time the design process concludes. With this in mind, it is crucial that companies make correct sourcing decisions as early as possible.

Suppliers can influence systems technology that directly relates to a product's function, product structure, and cost structure and ultimately to customer requirements. To understand product costs and ensure production flexibility, Apple's highly visible work on its iPod product illustrates a successful example of strategic supplier interaction. One of the areas in which the iPod changed the competitive landscape was its coupling with iTunes and ultimately with the entertainment companies that supply music. This combination of hardware, software, and content represents a unique value chain. This supplier relationship not only ensures the availability of content for users but also facilitates the strategic sourcing of hard drives and flash memory.

Other successful examples of closely managed supplier relationships include IBM, with its integrated product and development processes; Proctor & Gamble, with its internal connect and development strategies; and Dell, which has become the poster child for supply chain management.

Lower Total Cost of Ownership and Life Cycle Costs

Total cost of ownership includes all the costs of acquiring, owning, or converting an item of material, piece of equipment, or service and post-ownership costs.[18] This includes the disposal of hazardous and other manufacturing waste and the cost of lost sales as a result of a reputation for poor product quality caused by defective materials or purchased services that are incorporated in the end product or services. Purchasing can influence the costs in many areas, including product design through early involvement, acquisition, and risk costs through effective supplier selection decisions. Purchasing can also influence cycle time, conversion, supply chain costs, and even post-ownership costs.

Working with suppliers (or potential suppliers) and recognizing the total cost elements can also help in negotiations. There are often creative ways to reduce the items that are a large percentage of the total cost. Also, purchasing is taking a more strategic role in identifying opportunities to reduce the environmental footprint of a product over its life cycle. These opportunities help organizations meet the ever-increasing regulations and also help to build brand and reputation for the consumer.

Case in Point—Purchasing's Role in Reducing Carbon Footprint

Both saving money and reducing the carbon footprint across a supply chain are responsibilities of purchasers. There are many ways that purchasing can be involved in reducing the carbon footprint. The first is understanding what drives carbon output. This requires gathering data, both internally and externally from the suppliers. This is often the most challenging step, and many organizations are gaining expertise, such as the Carbon Disclosure Project.[19] According to the CDP, 10 percent of the world's largest companies are generating 73 percent of all greenhouse gases (GHG).

After data gathering, the next step is to look at the largest drivers of greenhouse gas emissions and measure the life cycle costs and impact. A developing technology is making this task easier for purchasers. The Carbon Trust is one such organization.[20] The Sustainability Consortium, founded by Wal-Mart, is another organization that has a standard platform for companies to measure carbon footprint.[21]

There are many drivers of GHG, but two of the primary drivers are transportation—both distance and mode—and customer use of the products. Purchasers can influence both of these drivers by considering supplier location, thinking about the materials that go in the products, and thinking about the packaging of the products. If an item needs to be returned, what is the cost of doing this? Are products designed for easy assembly and disassembly? Does it make sense to source from a low-labor-cost area, pay increased transportation costs, and increase carbon outputs? These are some trade-offs that need to be considered (and are discussed in later chapters).

The key is to communicate environmental goals to suppliers to ensure that they focus on energy performance and environmental performance of their products and services. Incorporating metrics into the supplier scorecard that evaluate environmental and energy performance can start the process. Purchasing can be involved in the reduction of total costs, life cycle costs, and therefore carbon in many ways.

Purchasing and Supply Management and Return on Investment (ROI)

Every dollar saved in purchasing is equivalent to a dollar of new income. Because purchasing is responsible for spending more than one-half of most companies' total dollars highlights the importance of purchasing's contribution to the bottom line. Figure 1-2 shows the relationship between cost-savings and ROI. This model is often called the Dupont model and is used in many situations to show how material cost-savings have a greater influence on the ROI than do sales increases, or how it is easier for companies to generate cost-savings versus increasing market share. With trained and skilled purchasing personnel, these cost-savings are often much less difficult to achieve than significant increases in market share. As indicated earlier, purchasing can influence both the top and bottom line of the organization. These two working in tandem have a significant influence on shareholder happiness.

Figure 1-2 shows what happens if materials costs are decreased by 5 percent, or $120,000. The numbers in red are after the reduction. Given this scenario, the value of inventory also decreases 5 percent, but this is not following standard account rules and assumes that the value of the inventory is zero when the products are received at the reduced material costs. A cost-savings, in this situation, of $120,000 increases ROI by more than 3 percent. This is a valuable presentation to show the strategic opportunity that supply management has in improving stakeholder value.

Purchasing's Role in Business

Purchasing is one of the basic functions common to all types of business enterprises. Business involves coordinating and integrating the six functions listed here,[22] which all fall under the basic plan, source, make, deliver supply chain model made popular in the late 1990s.[23]

1. **Creation**—The idea or design function
2. **Finance**—The capital acquisition, financial planning, and control function
3. **Personnel**—The human resources and labor relations function
4. **Supply**—The acquisition of required materials, services, and equipment
5. **Conversion**—The transformation of materials into economic goods and services
6. **Distribution**—The marketing and selling of goods and services produced

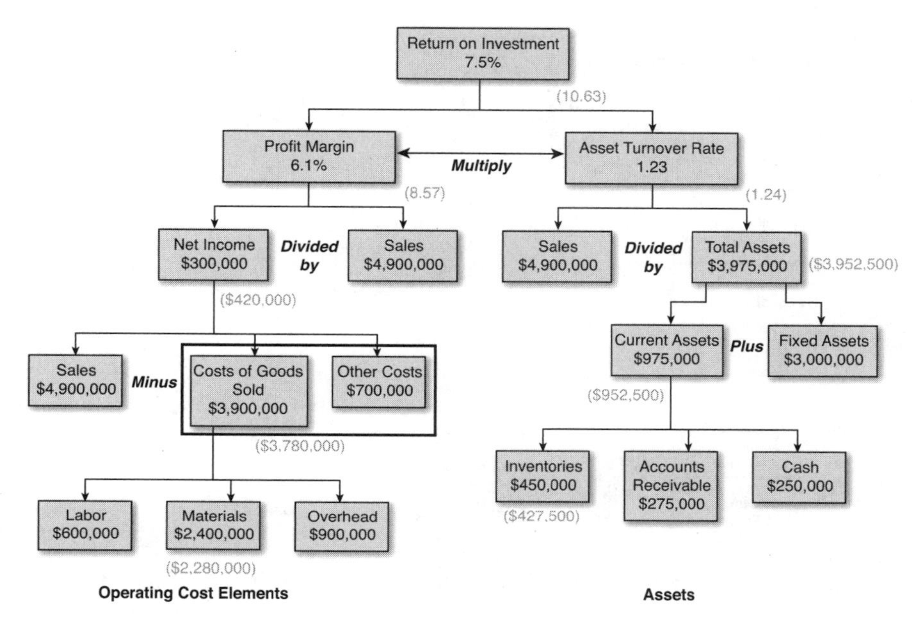

Figure 1-2 Supply management and ROI

A number of organizational units are responsible for executing the six functions. For example, research and design are typically engineering functions and most likely involved in the creation aspect. There are finance and accounting departments that typically manage the flow of financial resources into and out of the firm. The purchasing department is responsible for supply—however, other functions are involved as well, such as marketing in the purchase of advertising services.

Supply management has many interfaces with the different organizational units responsible for executing the primary business functions. Figure 1-3 represents many of the internal interfaces of the supply management function and just some of the many activities when the functions are required to interface. These internal organizations represent many of purchasing's customers. Purchasing has a wide intra- and inter-organizational footprint.

Many other areas of involvement exist within the organization that are not included in Figure 1-3, and, of course, many additional activities where the functions in Figure 1-3 interact. Many of these activities are discussed throughout this text. Some of the other interactions are briefly discussed here.

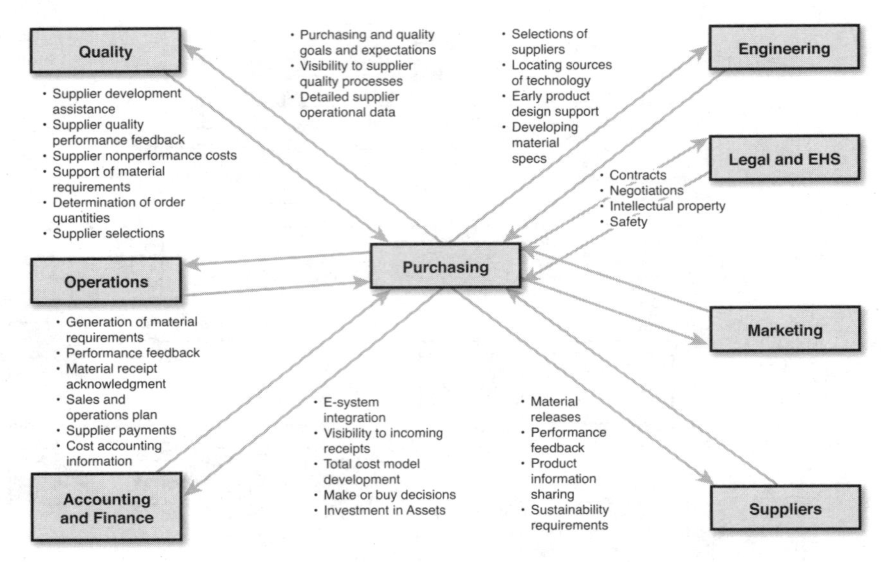

Figure 1-3 Purchasing's intra-organizational relationships and activities[24]

Purchasing and Engineering

The product costs associated with quality, material, fabrication, and production are linked to the design specifications. Specifications can be written in a manner that reduces or enlarges the number of firms willing to supply specific items. There can be conflict between engineering and purchasing simply because engineering tries to design the "ideal" product or services without regard for cost or availability of resources.

Often, differing performance metrics between purchasing and engineering generate the conflict, and many times this conflict is not easily resolved. However, involvement of purchasing in initial design conversations with engineering may facilitate better discussion and ultimately results. Another service that purchasing can provide engineering is helping to use "like components" so that there is no need to develop a new component that is already purchased for a different product line. Purchasing can help in reducing SKU proliferation, a problem common in many organizations.

One example of the relationship between engineering and purchasing is the case of a global electronics manufacturer. The specifications for pallet dimensions were supplied to the suppliers, but the suppliers were not following the specification and instead were shipping on multiple-sized pallets in varying conditions. The purchaser held the suppliers accountable for the appropriate pallet dimension and quality, and considered each one sent incorrectly as a defect. After standardization, damage from shipping decreased significantly, and the volume shipped per truckload increased significantly. This particular project led to a number of other improvements in packaging, waste reduction, improved container and trucking capacity utilization, and much more.

A company in the beauty products industry was delivering products to hundreds of countries around the world, many with different language and labeling requirements. In one instance, the company purchased a common jar for one of its products but then had the supplier label and ship each separately. This generated issues with SKU proliferation, and the company saw increasing inventories. The customers in China had different preferences than those in India and products would fall out of favor in one region before another. The purchasers worked with the engineers and the suppliers, and decided to "postpone" the labeling of the bottles. The volume of standard-size bottles increased significantly, the price of the bottles decreased, and the labels were added only when customer (and regional) specific demand was known.

Purchasing and Manufacturing

The timing and quantity of the receipt of purchases often strain the relationship between manufacturing and purchasing. Also, poor planning of requirements at the strategic level causes a bullwhip effect (discussed in the marketing interface). Purchasing needs sufficient time to qualify suppliers, develop competition, and negotiate and ultimately reduce opportunity for special production or premium transportation.

Trouble in this interface often arises because of poor forecasting and therefore poor production planning. Integrating demand and supply sides of the business in sales and operations planning has a tendency to improve the relationship and the outcomes. Also, coordinating in the early design stages for new products can help alleviate this conflict. New product development meetings are generally cross-functional and engage many members of the organization.

There are cases of standardization of materials, standardization of components, and simply reviewing material specifications prior to a purchase that have saved significant money over time. Also, in some cases it makes sense to pay more for an alternative material, as long as it is feasible to do so, and it may save in manufacturing costs in the long run.

Purchasing has to assist in achieving faster time to market and reduced time for changeovers and tool and line setup work by working with suppliers to improve capabilities and increase response time. Manufacturing has the goal of achieving faster time to market, decreasing operational costs and unnecessary setup times and waste. It is also often responsible for inventory costs and have a vested interest in keeping them at optimal levels. Shutting down a production line is extremely costly. There are reports of parts being helicoptered to an automobile manufacturing facility because the cost of a plant shut down was more than $1,000,000 per day. (That is what was invoiced to the supplier.) There are many stories of purchasers "flying" components to customers to avoid the penalties associated with shutting down a facility. Purchasers have had to find emergency sources of supply, often paying much higher costs to avoid a plant shutdown.

Purchasing and Marketing

Many marketing departments spend significant amounts of money on advertising and promotion. This is what typically generates sales. The problem is that sales and marketing activities are often not linked to supply and production activities. Customers often do not communicate promotions and therefore create even more supply chain problems.

Hau Lee, a professor at Stanford, coined the term *bullwhip effect.*[25] The bullwhip effect occurs as even small increases in demand prompt the "whip" to get ever larger as it progresses down the supply chain. The bullwhip is shown in Figure 1-4.

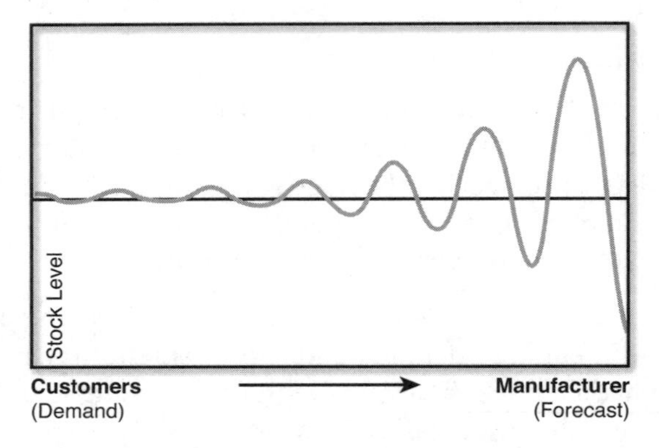

Figure 1-4 A visualization of the bullwhip effect

There are four major causes of the bullwhip effect, and much of the cause is generated in the downstream in the supply chain.[26] Upstream members tend to overcompensate because of historical issues with demand planning and forecasting.

- **Demand forecast updating**—The problem here is that as each entity in the supply chain updates the forecast, it also includes safety stock, tends to buffer the orders of others, and tries to replenish its own stock. An automobile company was having serious trouble with forecasting, and the relationship among sales, marketing, operations, purchasing, and its supply base was strained. Not one entity in the chain believed in the accuracy of other entities' "guesses." The result was that at each stage, the person responsible increased the forecast (or the plan) sometimes by as much as 10 percent. If one person said he needed 100, the next person in line would add another 10 to that order and so on. This process trickled down through the bill of materials and ultimately all inventories increased: finished goods, work-in-process, and raw materials. Suppliers increasingly had to expedite materials, and the entire supply chain was operating inefficiently.

- **Order batching**—Companies often place orders in batches. Sometimes, it is to reduce the administration of the orders, other times trying to out-guess the schedules of the manufacturers. In this situation, suppliers face erratic ordering with frequent increases and also frequent expediting because customers often don't want to wait for a batch run. This process was common in the furniture industry in which changeovers in manufacturing were expensive to manage. The products were manufactured in batches and the orders were often allocated to customers based on a set of criteria. Customers knew that if they over-ordered, they would likely receive more of the batch being manufactured.
- **Price fluctuation**—Special promotions and price discounts result in customers buying in larger quantities and stocking up. When prices increase, they stop buying, and the consumption patterns are destroyed. Retailers offer many pricing discounts and promotions to generate market demand. These promotions create demand that in effect will shortly vanish because customers will "stock up" while prices are low. An interesting story is about a family that bought peanut butter in bulk. When there was a promotion, and lower prices, the family would buy an entire case of the peanut butter. At one point, there was a salmonella scare in some of the brands of peanut butter. The "case" that was purchased was from the batches that were produced incorrectly. All the peanut butter was recalled and returned. All this fluctuation in demand patterns wreaks havoc on the supply chain. In addition, the forecasters will likely continue to use these demand patterns to forecast future demand and the problems continue to escalate.
- **Rationing and shortage gaming**—When supply is low, manufacturers may ration their products. Customers, in an effort to get "the most," will game the system by over-ordering. Generally, those customers with the most power will try to "hold" the product availability until their demands are known by placing unrealistic orders far in advance. When their demand is known, the orders are changed to reflect more realistic demand patterns. However, supplies were already ordered, labor was already scheduled, and capacity was already fixed on one product when it should have been on another.

All these bullwhip issues still occur today with potentially larger impacts on an extended supply chain. This is more than 15 years after the issues were identified. Prompt and frequent communication throughout the supply chain but especially from sales and marketing to manufacturing and purchasing about changes in sales forecast and expected changes in demand is necessary. Information sharing is one key to minimizing the bullwhip effect on the supply chain.

Another area in which purchasing and marketing are just beginning to interface is in applying the purchasing process to the marketing and advertising spend. Purchasers do not interfere on the design of these services but instead ensure contract compliance and are often involved directly in negotiations.

Purchasing and Finance

Finance is usually purchasing's best friend and biggest supporter. The reason for this is that purchasing tries to develop ways to save money; it works with the supply base to ensure that the contracts are followed; and it helps to minimize overbilling and underbilling. There are many reasons that purchasers are involved with finance.

Poor financial planning and execution are the major causes of business failure.[27] Purchasing is responsible for managing a lot of the financial resources available within a firm. There is a delicate balance between economic conditions and an organization's financial resources. In some cases, it makes sense to allocate the organization's finances to forward purchases to avoid a higher price. However, this is a decision that must be made carefully. There are many instances in which purchasers decided that forward buying made the most sense, but then prices fell and the purchaser was "stuck" with high-priced inventory.

Finance also has to be willing to pay suppliers in a timely manner. Delayed payment of invoices has serious implications on the buyer-supplier relationship and may also impact future pricing. Delaying payment to suppliers is a way to improve cash reserves; however, suppliers demand timely payments to obtain their own resources.

In later chapters, financial analysis both internally and externally on some key supply chain metrics help to formulate strategy and mitigate risk. For example, looking at a supplier's cash-to-cash flow can tell you whether they are effectively managing both payments and receipts. Managing a supplier's change in revenue compared with a change in costs can tell you how efficiently a supplier uses its resources. The same is true within your own organization. Keeping an eye on days payable and receivable outstanding can help minimize surprises.

As presented earlier, supply management has a major impact on ROI. Being efficient and effective in purchasing can significantly reduce the funds required to operate the firm. The timing and quantities of purchasing expenditures can significantly impact a firm's financial ratios.

Purchasing and Information Technology (IT)

In a later chapter, the use of technology to improve supply operations is discussed. Trending social network sites and other open-sourcing programs make it challenging sometimes to keep up with the many market changes. IT helps purchasing streamline processes, increase information, and obtain access to necessary data.

Many firms purchase business-to-business e-commerce buy-side software systems from firms such as Arriba, Commerce One, and PeopleSoft. These systems help to streamline the purchasing process and facilitate communication internally and externally. Many of these systems also include database technology that provides timely and accurate input to supply management for strategic planning and tactical activities. Chapter 4 is dedicated to supply chain technology, and types of e-sourcing tools and trends are discussed. The relationship between purchasing and IT is changing as quickly as the technology to effectively source products and services.

Purchasing and Logistics

Logistics is concerned with the movement of goods, sometimes both inbound and outbound, to the organization. Usually, it controls outbound only. The logistics professionals design and manage a firm's distribution system, which consists of warehouses, distribution points, and freight carriers. In some organizations, purchasing plays a dominant role in sourcing and pricing of logistics services. In others, the logistics department performs these services with little or no supply management involvement.

Generally, the sourcing and logistics roles are not linked. This is primarily because warehouse management and transportation management systems are generally not linked to e-sourcing tools. Also, the transportation aspect is performed at the plant or facility level, whereas sourcing is often centralized at a more corporate level. There is also a lot of price volatility in the transportation industry, which requires a different type of skillset to manage and to measure supplier performance.

The primary idea in the linkage between sourcing and logistics is that whoever does the sourcing should use professional purchasing practices. This is a growing area of opportunity for collaboration and cost-savings. Many organizations have set their sights on improving their transportation spend and, as mentioned earlier, reducing their carbon footprint in the transportation area.

Purchasing and Legal

Legal professionals are frequently involved in contract negotiations and contract formation. They review and approve contracts developed by the purchasing professional. However, there is often little collaboration on working with the firm's legal suppliers.

As with the logistics area, these intra-organizational relationships vary within the area of services spend as indicated in the following case in point. This variation is often due to budget ownership of the service. For example, marketing and advertising spend is usually owned by the marketing area. The involvement of purchasing is often nonexistent or minimal in an area of spend that is potentially a significant portion of the organization's overall expenses.

Case in Point— Intra-Organizational Relationships Surrounding the Legal Spend

In later sections, the idea of commodity segmentation is discussed. However, in this case, the purchaser added value in the area of legal spend by viewing legal spend as a category of spend with many different types of purchases contained within. For example, the area of legal spend contains litigation and court reporting. There is always an expense in discovery and legal witnesses. Purchasing went after a number of these categories and realized that they were just like commodities, with significant opportunity to reduce costs.

For example, in the area of court reporting, the legal firms (suppliers) used different organizations to perform this service. It was more about who was available on a given date and time and geographic location. The legal supplier hired the court reporter and then passed through the cost to the buying organization (plus a markup).

Purchasing did some research on the court reporting industry and discovered that there was a pool of large suppliers that could perform this service. Purchasing offered the court reporting firms larger volumes and by doing this achieved a lower price. The relationship with the court reporting firm was through purchasing at the buying organization. Improving this one piece of the legal process helped the purchasing organization show success, and for this it was provided additional opportunities to participate in the legal arena.

The Purchasing Process

Purchasing is a highly complex organizational process with objectives that reach far beyond the traditional belief that purchasing's primary role is to obtain goods and services in response to internal needs. The overall goals of purchasing can be categorized in five major sections: supply continuity, manage the sourcing process efficiently and effectively, develop supply base management, develop aligned goals with internal stakeholders, and develop integrated purchasing strategies that support organizational goals and objectives.[28]

Development of a strategic sourcing plan is driven by the recognition that tactical sourcing will not succeed in yielding a supply base that results in the benefits of collaborative relationships and strategic alliances. The basic steps in the strategic sourcing process follows:

1. Discover potential suppliers.
2. Evaluate potential suppliers.

3. Select suppliers.
4. Develop suppliers.
5. Manager supplier relationships.

Discovering Potential Suppliers

The ability to discover qualified and competent suppliers has increased exponentially with the introduction of the Internet. However, purchasers should not ignore other sources of information that are available to ensure that the supplier pool consists of the most appropriate suppliers whether domestic, nearshore, or offshore. Following is a list of resources to use in establishing a robust list of potential suppliers:

- Supplier websites
- Supplier information files
- Supplier catalogs
- Trade registers and directories
- Trade journals
- Phone directories
- Mail advertisements
- Sales personnel
- Trade shows
- Company personnel
- Other supply management departments
- Professional organizations

Other strategic issues to consider in determining the list of potential suppliers include

- Company policy on single versus multiple sourcing
- Company policy on buyer's share of supplier's capacity
- Company policy on buying from minority- and women-owned business enterprises
- Company policy on environmental, health, and safety (EHS)-qualified or certified suppliers

Evaluating Potential Suppliers

After developing a comprehensive list of potential suppliers, the supply manager's next step is to evaluate each supplier individually. The type of evaluation varies with the nature, criticality, complexity, and dollar value of the purchase to be made. In 1983, Peter Kraljic developed a matrix that helped to describe the type of relationship with the supplier according to the criteria mentioned previously.[29] Figure 1-5 is an adaptation of this matrix.

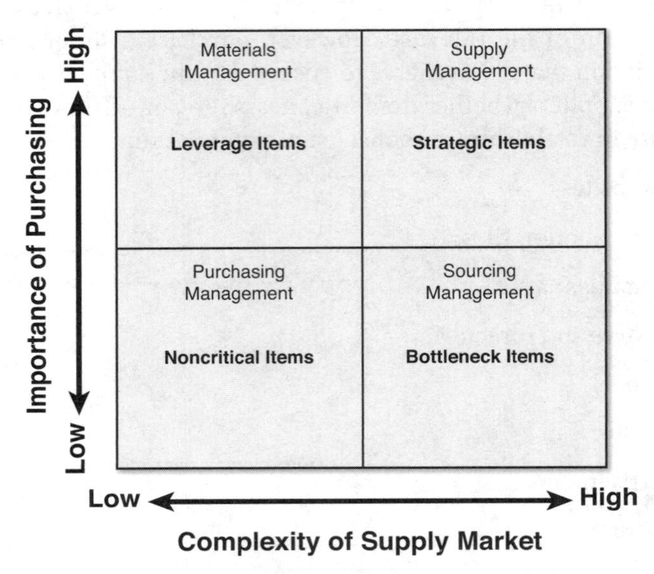

Figure 1-5 Kraljic's 1983 portfolio matrix

This step in the strategic sourcing process is key because there is a direct relationship between supplier relationship management and supplier performance, risk management, and brand and image. Purchasing's relationship with the supplier depends on the classification of the commodity in the matrix. The vertical axis is based on the importance of the purchase to the buying organization, usually talked about in terms of total spend. However, it can be high in importance because if you can't find the item, you cannot produce your product or service. Complexity is determined by the number of available suppliers. Fewer available suppliers make the purchase more complex. This matrix is discussed more in depth in Chapters 2, 3, and 4 and is used extensively by sourcing professionals.

Many of the uncomplicated, low-dollar-value items require only a cursory evaluation process because of the low importance and low risk to the firm. The role of the purchaser in this case is to buy at the lowest price. The idea is to streamline the buying process for these items. Locating suppliers for items in this category may include review of the supplier website or a look at a business database such as Mergent OnLine to gather relevant information.

In comparison, for complex or high-dollar or other critical purchases, the evaluation process is more involved, time-consuming, and costly. A key first step is to establish some knock-out criteria: In other words, what are some important concerns that if the supplier doesn't have these, they can't do business with your organization? Some examples include the size of facility, location of facility, past experience with similar requests, and litigation or EHS issues. There are a number of other evaluation techniques used to assess the suppliers, as shown in Table 1-4.

Table 1-4 Evaluation Techniques for Potential Suppliers[30]

Technique	Description
Supplier surveys	Surveys ask a number of questions of the supplier, including referrals, references, P&L history, defect rate, and quality management system.
Financial condition analysis	Financial analysis can often prevent the expense of further study. This analysis includes key financial metrics and ratios that assess the financial stability of the supplier. Credit ratings can also help determine whether the supplier can meet the demands.
Third-party evaluators	Trained third-party organizations are often hired to evaluate and audit suppliers or even processes like handling of hazardous waste.
Evaluation conference	Face-to-face discussion can help clarify specifications and determine whether a supplier can meet the demands of a complex purchase.
Facility visits	Many suppliers look good on paper, but visiting a site can help determine whether there are inefficiencies. These visits usually include a cross-functional team with both strategic and tactical participants from the buying firm. Weighted scorecards are often used during evaluation.
Quality capability analysis	The quality department and top management help to shape the quality capabilities of a firm. Understanding the supplier quality philosophy and past quality performance can help determine whether the supplier is in alignment with the buying company.

The supplier's strategies must also be aligned with the strategies of the buying organization. A supplier scorecard is usually cross-functionally developed with weights assigned to the different areas. (The scorecard and evaluative criteria are also discussed further in Chapter 2.) The final score on the scorecard helps to narrow the supplier pool but also allows the evaluator to focus on those things that are critically important to the buying organization.

The four focus areas listed (early supplier involvement, ethical considerations, environmental considerations, and social considerations) need additional consideration with each of the suppliers. It often depends on where the commodity or service is classified on the matrix, what type of relationship you have with the supplier, and where the supplier is geographically located because legal and cultural laws influence all these criteria.

Selecting Suppliers

When the supplier pool is reduced to a manageable level and one or more potential suppliers has passed the initial evaluation process, the purchasing manager or sourcing team can invite potential suppliers to submit bids or proposals. Purchasers have to decide whether to use bidding or negotiation, or some combination of both. Reverse auctions are often utilized here (depending on the classification).

As mentioned earlier, the final selection of suppliers is often based on a supplier scorecard or weighted factor analysis. Developing a weighted scorecard consists of four primary activities:

- Develop the factors that serve as the selection criteria and the weight that each of those factors carries in decision making. What areas are critical to your organization for this type of commodity or service? For example, a paper company has to buy significant quantities of starch for its manufacturing process. Starch is truly a commodity and the primary consideration is price.
- Expand the subfactors or performance factors within the broader selection criteria and the weighting of those factors. An example might be financial performance and key ratios of inventory turnover, return on assets, or even profitability.
- Establish a scoring factor to evaluate potential suppliers. This is generally the scale used for evaluation: 1 through 5, for example. The raters have to be clear on what a 1 is compared with a 5.
- Score and evaluate each supplier. This is generally done individually by those who have a relationship with the supplier and for those who have access to the information. For example, on-time delivery may not be known by all members, but the person responsible rates the supplier. The scores are all compiled and totaled to get to the scorecard "number."

Careful evaluation of the suppliers using the scorecard enables the appropriate selection of the supplier that clearly supports the needs of the buying organization. It is possible that the highest number is not the best supplier simply because the highest priority factor is what actually matters (for example, quality).

Developing Suppliers

Supplier development is any activity undertaken by a buyer to improve a supplier's performance or capabilities to meet the buyer's short- and long-term supply needs. There is sometimes conflict between the buying firm and the supplying firm, especially if a supplying firm does not see the need for post-development. Also, it is critical for purchasers to have a defined set of performance metrics that are transparent to the supplier and established goals for development. Effective supplier development requires the commitment of financial capital and skilled personnel, timely and accurate information sharing, and process improvement. More detail on supplier development is provided in later chapters.

Managing Suppliers

Key performance metrics are in place to help manage the suppliers. However, purchasers must assess the supplier's capabilities to meet the firm's long-term needs. The buyer must be willing to ask the supplier about general growth plans, design capabilities, and future production capacity. An important part of managing the supplier is building and maintaining the appropriate relationship. More detail on supplier management is provided in later chapters.

Strategic and Tactical Roles of Purchasing

There are a number of key strategic roles and tactical responsibilities for purchasing embedded in each of the process steps. Purchasing's "span of control" gives them final authority over certain matters. However, the internal customers (or budget owners) have a strong influence over many important decisions. Some of these key strategic roles and tactical responsibilities for purchasing are listed here. All these are discussed in greater detail in later chapters.

Strategic Roles of a Purchaser

- Spend Analysis
- Demand Management and Specifications (Statement of Work)
- Category Management
- Contract Management
- Cost Management
- Managing and Improving the Procure to Pay Process
- Supplier Relationship Management
- Establish a Supply Management Strategy

Tactical Responsibilities of a Purchaser

- Supplier Identification/Evaluation/Selection
- Forecast and Plan Requirements
- Needs Clarification: Requisitioning
- Purchase Requisitions/Statement of Work
- Review Forecasts and Customer Orders
- Establish a Reorder Point System
- Stock Checks (Cycle Counts)

Types of Purchases

Organizations buy many different goods and services. As previously indicated, the challenge for purchasing is deciding on the supplier that offers the best opportunity for items an organization must purchase externally. Table 1-5 lists and describes many of the items that a purchasing department is responsible for buying. Services are a special category of spend and the involvement of purchasing depends on the organization.

Table 1-5 Different Types of Purchases

Type of Purchase	Description	Examples
Raw materials	Items with a lack of processing by the supplier into a newly formed product. Often these raw materials are not of equal quality and are purchased by "grade."	Petroleum, coal, lumber, copper, zinc, gold, and silver
Semi-finished products and components	All items purchased from a supplier required to support an organization's final that are production.	Components, subassemblies, assemblies, subsystems, and systems (seat assembly, steering assembly, doors, and posts)
Finished products	Products for internal use or products that require no major processing before resale to the end customer.	Furniture, computers, cars, and carts
Maintenance, repair, and operating items (MRO)	Items that do not go directly into an organization's product but are required to run the business.	Spare parts, office and cleaning supplies

Type of Purchase	Description	Examples
Production support items	Materials required for packaging and shipping.	Tape, bags, inserts, and shrink-wrap
Services	Services required to support the facility or the business.	Customer support, temporary labor, facilities, and legal
Capital equipment	Assets intended to be used for more than one year.	Machinery, computer systems, and material-handling equipment
Transportation and third-party purchasing	A specialized type of service buying to manage inbound and outbound material flows.	Rail, truck, ocean, 3PL, and multimodal

Conclusion and Chapter Wrap-Up

The goal of this chapter is to discuss the roles and objectives of purchasing within an organization and in supply chain operations. This chapter describes purchasing's role in the organization and the interaction with other organizational functions. There is much discussion on different ways that purchasing can add value, including involvement in new product development. The bullwhip effect is introduced to show the many ways that conflict and complexity are driven into the supply chain.

The strategic levels of the purchasing process are touched on, and are discussed in much more detail in later chapters. A number of different types of products and services that purchasing buys are noted. Key terms used throughout the rest of the book are also defined throughout this chapter.

The key points from this chapter include

- Understanding the important role of purchasing in an organization
- Assessing the purpose and goals of the supply management organization and its contribution to profitability and the bottom line
- Identifying the relationship between the purchasing function and other functional areas
- Identifying the activities that are part of supply management
- Understanding the different categories of purchasing spend

Key Terms

- Purchasing
- Supply chain management
- Procurement
- Strategic sourcing
- Price
- Total cost of ownership
- Life-cycle cost
- Outsourcing
- Offshoring
- Supplier development
- Core competency
- Procure to pay
- Span of control

References

1. Monczka, et al. (2011). *Purchasing and Supply Management*, 5th Edition. South Western Publishing: Mason, OH.
2. Rawat, N. (2009). "Purchasing and Supply Chain: History of Purchasing.," Retrieved September 27, 2013, from http://navpurchasing.blogspot.com/2009/07/history-of-purchasing.html.
3. Ellram, Tate, and Billington (2008). "Services supply management: the next frontier." California Management Review.
4. Bureau of Economic Analysis (2013). U.S. GDP. Retrieved on September 27, 2013, from http://www.bea.gov/itable/index.cfm.
5. Fine, C. (1999). Clockspeed.
6. Ellram, Tate, Billington, ibid.
7. Benton, W.C., Jr. (2007). *Purchasing and Supply Management*. McGraw-Hill: New York.

8. Spekman, et al. (1999). "Toward More Effective Sourcing and Supplier Management. European Journal of Purchasing and Supply Management, 5 (2), pp. 103–116.

9. Eisto, et al. (2010). "Early Supplier Involvement in New Product Development." *World Academy of Science*, 38. Retrieved on September 23, 2013, from http://www.waset.org/journals/waset/v38/v38-159.pdf.

10. Monczka, et al. (2011). *Purchasing and Supply Management*, 5th Edition. South Western Publishing: Mason, OH.

11. Monczka, et al. (2011), ibid, p. 217.

12. U.S. Food and Drug Administration (FDA). "Animal & Veterinary, recalls and withdrawals." Retrieved on September 15, 2013, from http://www.fda.gov/AnimalVeterinary/SafetyHealth/RecallsWithdrawals/default.htm.

13. FDA, (2013). Retrieved on September 15, 2013, from http://www.fda.gov/Safety/Recalls/ucm367257.htm.

14. CBS, (2013). "P&G Recalls Dry Pet Food over possible Salmonella Contamination." Retrieved September 15, 2013, from http://newyork.cbslocal.com/2013/08/15/procter-gamble-recalls-dry-pet-food-over-possible-salmonella-contamination/.

15. 1000 Ventures (2013). "How to win market share." Retrieved on September 23, 2013, from http://www.1000ventures.com/business_guide/market_leader.html.

16. Siemens (2013)." Leveraging suppliers for strategic innovation." Retrieved September 27, 2103, from http://www.plm.automation.siemens.com/zh_cn/Images/7562_tcm78-4602.pdf.

17. Brown, J. (2005). "Procurement in New Product Development," The Aberdeen Group, Inc.

18. Ellram, L.E. Total cost of ownership.

19. Carbon Disclosure Project (2013). "Driving Sustainable Economies." Retrieved on September 15, 2013, from https://www.cdproject.net/en-US/Pages/HomePage.aspx.

20. Carbon Trust. (2013). "Footprint Measurement." Retrieved September 15, 2013, from http://www.carbontrust.com/client-services/footprinting/footprint-measurement.

21. Sustainability Consortium (2013). Retrieved on September 15, 2013, from http://www.sustainabilityconsortium.org/.

22. Monczka, et al. (2011), ibid.

23. Supply Chain Organization (2013). Supply Chain Operations References (SCOR) Model. Retrieved on September 15, 2013, on http://supply-chain.org/f/SCOR-Overview-Web.pdf.

24. This figure is adapted from the work of Monzcka, et al. (2011); Burt, Dobler, and Starling (2003), "World Class Supply Management." McGraw-Hill: New York; W.C. Benton, Jr. (2007) *Purchasing and Supply Management*, McGraw-Hill: New York.

25. Lee, H., et al. (1997). "Information Distortion in Supply Chains," *Management Science*, 43 (4), pp. 546–558 and Lee, H., et al., (1997), "The Bullwhip Affect in Supply Chains," Sloan Management Review, 38 (3), pp. 93–102.

26. Ibid.

27. All Business Experts (2013). "The Top Ten Reasons Why Businesses Fail." Retrieved September 24, 2013, from http://experts.allbusiness.com/the-top-10-reasons-why-small-businesses-fail/889/#.UkFu-yuDR7M.

28. Burt, Dobler, Starling (2005), ibid.

29. Kraljic, P. (1983). "Purchasing Must Become Supply Management." *Harvard Business Review*, September–October, pp. 109–117.

30. Monczka, et al (2011), i. Ibid.

2

KEY ELEMENTS AND PROCESSES IN MANAGING SUPPLY OPERATIONS AND HOW THEY INTERACT

This chapter brings together the key processes involved in managing the many activities and procedures involved in supply operations. Category analysis, introduced in Chapter 1, is discussed in more depth in this chapter. Also, some of the more tactical aspects of the supplier selection and evaluation process are detailed. Many of the concepts introduced in Chapter 1 are expanded upon in this chapter. Also, the idea of maintaining and developing suppliers along with effective performance evaluation techniques are introduced. The key concepts of all these processes are defined throughout the chapter.

Learning Objectives

After completing this chapter, you should be able to:

- Understand the basic processes involved in managing supply operations.
- Determine how to establish a commodity strategy and perform a category analysis.
- Perform a supplier screening and selection.
- Understand how to negotiate and manage contracts.
- Validate supplier performance and quality.
- Develop the appropriate supplier relationships and development of suppliers.

Introduction to Supply Management Processes and Commodity Strategy Development

The competitive marketplace demands that firms relate to and react to changes in the environment. To be effective, these firms must anticipate changes, adjust to changes, and capitalize on opportunities by formulating and executing strategic plans.[1] Purchasing strategies have to be designed to permit the achievement of specific organizational goals and objectives.

Purchasing and Risk Management

Purchasers must execute those strategic plans in a way that protects the organization from operational, financial, and reputational risk (Figure 2-1). The role of purchasing in risk mitigation, financial, reputational, and operational risks are defined and discussed next.

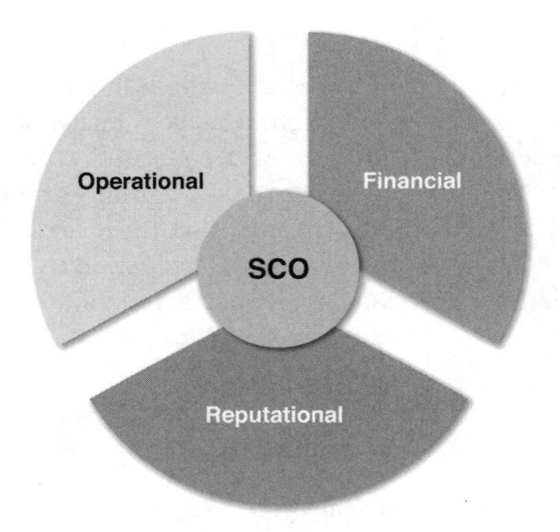

Figure 2-1 Three categories of supply operations risk

Financial risk looks at changes in price and the cost of materials and services. Purchasers must also be particularly sensitive to the financial stability of their suppliers. Constant monitoring of volatile commodity markets, attention to labor costs, and unionization that may put some suppliers in financial peril are critical to try to avoid potentially disruptive problems. As mentioned in discussions with the purchasing-finance interface, purchasers also have to watch accounts payables to ensure that discounts are not lost or the payables to suppliers are unnecessarily delayed. If suppliers don't get paid, they can't pay their bills. This might inhibit their receipt of raw materials.

Purchasing has significant opportunity to help to manage and mitigate supply chain financial risk. Ensuring that suppliers do not get into a situation in which they might go out of business is one way. Also, purchasers can help avoid costs of bad quality or harm to the firm's reputation and brand.

Purchasing in the purchasing function must be diligent in ensuring supplier viability and must have strong financial acumen. Even requests for price increases from suppliers for products or services can be reduced or negated by the involvement of purchasing. Many techniques that can be used in financial management of suppliers are discussed in later chapters.

Case in Point —Financial Risk and Evaluation

During the economic downturn in 2008, suppliers were hit particularly hard as organizations "tightened their belts." There was also a tendency for organizations to extend their days' payable to ensure that they had cash for operations. Purchasing was quickly moving from annual evaluations of their suppliers to weekly in some cases, checking to ensure that the suppliers were viable for the long term. In some cases, the suppliers could not outlast the economic problems and were forced into bankruptcy. However, some proactive companies kept a close watch on the performance of the suppliers, realized there was a potential problem, and called to find out what they could do to help. In some cases it required a short-term investment or potentially some equipment or tooling. The more proactive companies realized how important the suppliers were to the firm and knew the impact of a disruption on the supply chain. These organizations invested in the short term in their suppliers so that in the long term they would still have the relationship.

Reputational risk can have the most serious influence on an organization. This area of risk relates to both legal and ethical supply issues. It is an underlying assumption that "you are known by the company you keep." This applies both personally and to businesses. In other words, what your supply chain partners are doing has a direct impact on how your organization is perceived.

Many stakeholders see suppliers as an extension of the company conducting business. As such, any supplier—large or small—can tarnish a company's reputation and diminish brand value.[2] This is becoming increasingly problematic in an era in which many products and services come from "emerging" or "low-labor" cost locations. Supply chains are extremely large and complex. To manage this ongoing brand and reputational risk, companies are implementing supplier qualification programs that ensure alignment of the supplier's values and goals with those of the buying organization.

Purchasers don't want to have to stand in front of the news cameras to explain why one of their actions caused serious injury or even death to its customers or supply chain members. Adverse publicity for bribery, kickbacks, improper quality, improper disposal and environmental practices, dealings with unethical suppliers, and so on can be extremely damaging and costly.

Case in Point—Reputational Risk

There are a number of highly publicized cases in which the selection of the supplier caused reputational harm to the buying organization. Companies such as Mattel and Hasbro were purchasing toys that contained lead, causing harm to children. Apple and other electronics manufacturers were purchasing components from FoxConn, which had poor relations with its employees. Clothing retailers and distributors were sourcing from suppliers located in Bangladesh. These suppliers failed to follow appropriate safety procedures leading to deaths of many employees. In all these situations, it was the actions of the suppliers that created negative publicity for the buying organizations.

Instances also exist in smaller organizations where bad purchasing decisions generated negative publicity and severe penalties. In one such case, a business owner tasked its purchasing department to find a supplier to dispose of five containers of hazardous materials that had been accumulating at the facility. These materials required safe handling and needed to be taken to the appropriate disposal facility. Using the Internet, the purchasing agent found a company that guaranteed the lowest price and without further verification contracted with the organization. The disposal company arrived on time and picked up five containers of chemicals and loaded them on its truck. On the way to the disposal site, the truck pulled off the side of the road and threw out all five containers into a public area, where some of them immediately started leaking the chemicals. The name of the buying company was noted on the containers. Both the police and the news crew contacted the buying company to try to understand what happened and who was responsible. Who is responsible for this problem? Significant fines were levied, and there also is the possibility of a jail sentence.

Operational risk is associated with supply interruptions and delays. This type of risk arises from the people, systems, and processes through which a company operates. Some specific types of operational risks are considered important by organizational leaders; these are things that "keep them up at night." Purchasers need to identify and be well prepared for the potential of negative actions both internally and by supply chain partners to minimize the negative impact on supply chain operations.[3] Some of the potential high-impact supply chain operational risks are discussed in the following paragraphs.

- **IT sabotage**—There have been a number of instances in which IT systems are infiltrated to gain access to data. Chapter 4 discusses other IT issues such as cloud computing that open up more opportunity for potential problems. Purchasers have to ensure that contracts clearly stipulate access levels and information available to supply chain partners. Data redundancy should be built in to the system. Sensitive information such as pricing and costing information should be handled differently than some of the other information such as data required for transportation and delivery of products.[4]
- **Compensation and incentives**—This is a highly litigious society that we live in and product and service liability issues can result in claims that cost millions of dollars. Purchasers must carefully select and manage the supply base to help minimize these issues. Carefully screened and fully tested products should be shipped on time and complete, where and when needed.
- **Fraud and customer data abuse**—Fraud is a serious issue in times of economic downturn because of anticipated financial pressure. Purchasers must follow stringent guidelines and act ethically in all aspect of business dealings. They must always do what is in the best interest of the organization, not what is in their own best interests.
- **Epidemic disease**—A disease that impacts a specific region of the world may have serious implications to delivery. If a supplier's workforce is suddenly reduced because of illness or death, it can be difficult to find another supplier to fill the need. Many organizations have a sourcing policy that helps spread this risk by having multiple suppliers, generally located in different areas of the world.
- **Political intervention, regulation, and unrest**—Many concerns relate to currency issues such as the breakup of the Eurozone.[5] Many issues like that occurred in Barcelona, Spain, where there was an uprising over the lack of available jobs. Political events can occur suddenly, without warning. Duties, tariffs, taxes, and boycotts may change and have serious implications for the cost of doing business in a region.
- **Business continuity and disaster recovery** —It is difficult to plan for events such as Hurricane Sandy that seriously damaged the eastern seaboard in the United States in 2012. Maintaining business flow and recuperating from serious disasters will continue to be a challenge. As the weather patterns change and global warming causes many of the large-scale events, supply managers have to be more prepared and to recover more quickly than ever before. Dual sourcing of products from different regions of the world, carrying additional inventory during the more volatile seasons, or thinking about alternative transportation channels may help to alleviate this risk.

Managing supply risks can be challenging. However, designing, coordinating, and managing the operations of the supply system in a way that takes these risk concerns into account may provide a significant barrier against them. An organization can take specific steps to help reduce supply chain risk.[6]

1. **Get some visibility and think long term**. This involves understanding where the vulnerabilities are in the supply chain. Until you realize where things can break, you can't get a clear understanding of the long-term implications and how to fix the issues.
2. **Scenario plan failures**. Contingency planning is critical. Some risks are short-lived, and with some innovative thinking, they can be dealt with (that is, the 3-day no-fly zone over western Europe because of volcanic ash[7]); whereas others are more drastic and can create significant and long-term harm to the company (that is, scarcity of materials or lack of materials).
3. **Be realistic with customers**. Take a step back and think about each customer in terms of what it costs to serve the customer. In some cases, the largest customers are the least profitable (and can be negatively profitable). Be careful what you promise to the customer, and make sure that all the organization's functions are in agreement with those promises. The bullwhip effect discussed in Chapter 1 relies on the information and communication with customers and other supply chain members.
4. **Financial and credit assessment**. Even key suppliers should be routinely assessed for credit worthiness and or financial concerns. The viability of the supplier becomes clear just by looking at a few key metrics (discussed in later chapters).
5. **Have a plan**. The supply chain processes have to be clear and strictly adhered to. In the sourcing area, there are strict guidelines on how to find suppliers and how to measure and manage suppliers.
6. **Communicate**. One of the easiest ways to manage risk is to communicate your requirements as far in advance as possible. Talk to you suppliers; let them know they are important to you; and make sure that they understand your expectations.

Category and Portfolio Analysis

As you can see, managing risk is an important aspect of a supply manager's job and a way that purchasers add significant value to the organization. Chapter 1 discussed some other ways in which purchasing can add value to its shareholders, which involve increasing revenues and decreasing costs. To contribute and focus these efforts, the goals of supply management must be aligned with corporate objectives; in other words, supply management derives its strategic direction from these corporate objectives and business unit strategy development.[8]

The goals and strategies of supply management are often applied to particular categories, or general families, of purchased products or services. These categories may belong to a particular business unit that is then responsible for the actual performance of that particular classification. For example, office supplies are a common classification for all industries and are relatively unimportant to the success of the organization. Alternatively, lumber would be important and therefore has a more strategic classification in the pulp and paper industry. Reduced costs and improved customer service are often strategic initiatives across the different product classifications, which are greatly facilitated by the supply chain.

Managing these categories of products in a way that helps an organization to meet its organizational goals and strategically reduce the number of risk factors introduced into the supply chain is known as *commodity strategy development*. The sourcing process begins with an assessment of what you have, what you need, and what happens in the external market. Based on the established goals and objectives, the sourcing process might be initiated because new products are introduced, engineering changes are made to existing products, new technology is developing that you need access to, forecast and demand changes are occurring, or even new facilities are built. There are many reasons for the sourcing process to initiate. It is a matter of identifying a need for a particular item or service.

Spend and Demand Assessment—Procure to Pay

Following the category analysis is spend and demand assessment. Often gathering data on the total spend for a particular commodity is difficult. This is why many organizations promote and implement a *procure to pay process*. Procure to pay processes enable the integration of the purchasing department with the accounts payable department. Figure 2.2 shows a procure to pay process flow map. Procure to pay processes are designed to provide visibility and control over the entire life cycle of a transaction—from the way the item is ordered to the way that the final invoice is processed. This provides full insight into cash flow and financial commitments and reduces many of the inefficiencies inherent in the area of accounts payable.[9]

Some companies have the accounts payable function report up through the organization in the same channel as purchasing. Actually, this idea of an effective procure to pay process is at the top of many organizations' reengineering efforts.

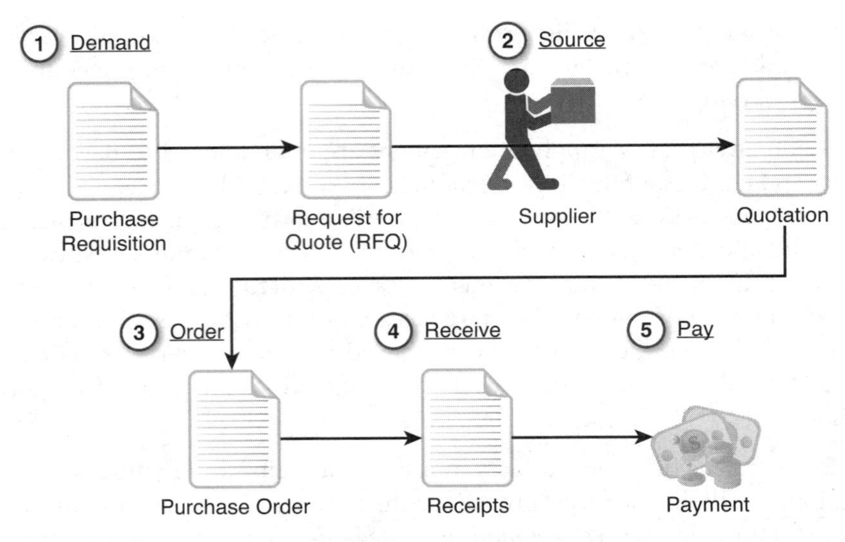

Figure 2-2 Procure to pay process flow[10]

Spend and Demand Assessment—Spend Analysis

The idea behind a spend analysis is to perform an annual review of the firm's entire set of purchases. This review looks at what the money was spent on and receiving of the right amount of products and services given what it paid for them. A spend analysis helps to determine the amount of spend per supplier, and if there is opportunity to consolidate volumes. The review also helps determine how actual spend compares to budgeted spend.

When the data for the spend is available, decisions can be made on how to move forward with the supply base and how to better link the commodity strategies with organizational goals.[11] The spend analysis helps the purchaser to identify the commodities in terms of importance to the organization and helps to further classify commodities and determine the type of relationship you have with the suppliers.

The two first steps in this spend analysis process involve sorting the commodities by category and then by supplier. This provides a visualization of where the assessment can have the most impact by looking at the categories of the greatest spend, and the suppliers who received the greatest about of business. Visualizing the spend in this fashion helps to mitigate some potential risk and also helps to identify the areas that offer the largest opportunities for cost-savings.[12] Figures 2-3 and 2-4 show examples of these visualizations.

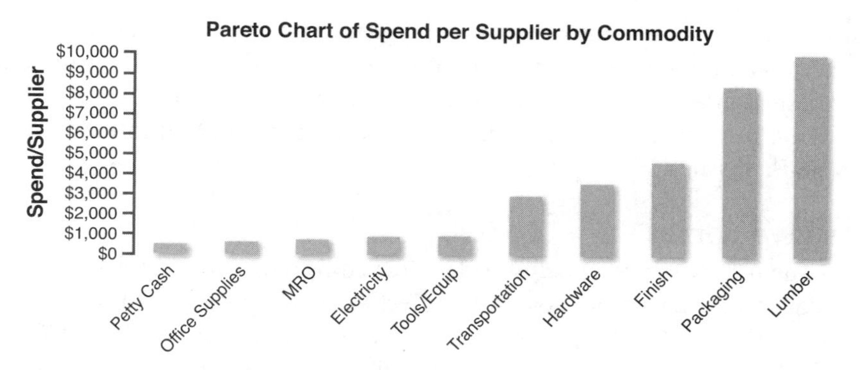

Figure 2-3 Commodity category by volume of spend

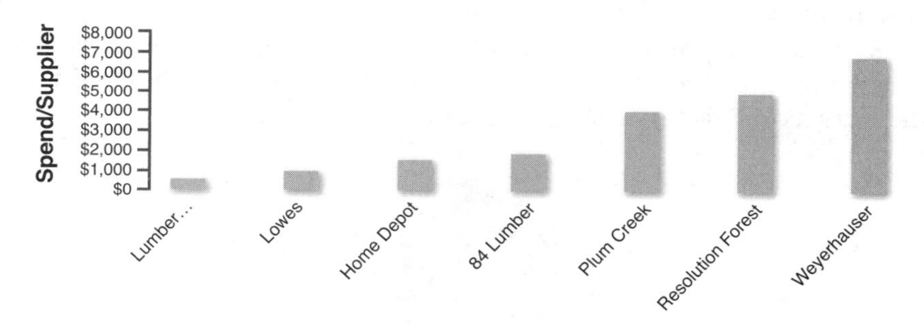

Figure 2-4 Commodity spend by supplier for lumber

Conduct Market Research

The next step is to drill down from the spend analysis into the specific commodities that the purchaser is responsible for analyzing. Look at the total amount of spend on the specific commodity compared to the total spend, and the suppliers that you spend the money on for the commodity. Generally, only a few suppliers receive a significant volumes or level of spend for a particular commodity. In many cases, this is the first step in optimizing or streamlining the supply base: understanding which suppliers are key and eliminating those that aren't making a contribution.

There is some key information that can help a purchaser make an informed decision. For example, information on total annual purchases is important by supplier and by business unit. In addition, forecast requirements from stakeholders and new sourcing elements should be included in decision making, engineering changes, new product launches, and others.

External market research is also key in identifying information about suppliers, available capacity, technology trends, pricing and cost data and trends, environmental and regulatory issues, and any other issues. Making an informed decision requires quite a bit of information gathering, education on market conditions, and an intense scanning of other key market issues.

Porter's Five Forces

Next, the issue is how to present all this data in a comprehensive and thorough way that understandably links corporate goals and objectives with specific commodity strategies and objectives. One tool that is key is Michael Porter's Five Forces Model (1980).[13] Porter introduced a framework that helps predict supplier and buyer behavior and is critical in helping to formulate supply strategy. This model also helps other stakeholders understand current market conditions. See Figure 2-5 for a presentation of the Five Forces Model. The factors that influence each of the constructs are discussed next along with tools on how this model can be used from the perspective of a purchaser to thoroughly assess the markets and factors that influence each of its constructs.

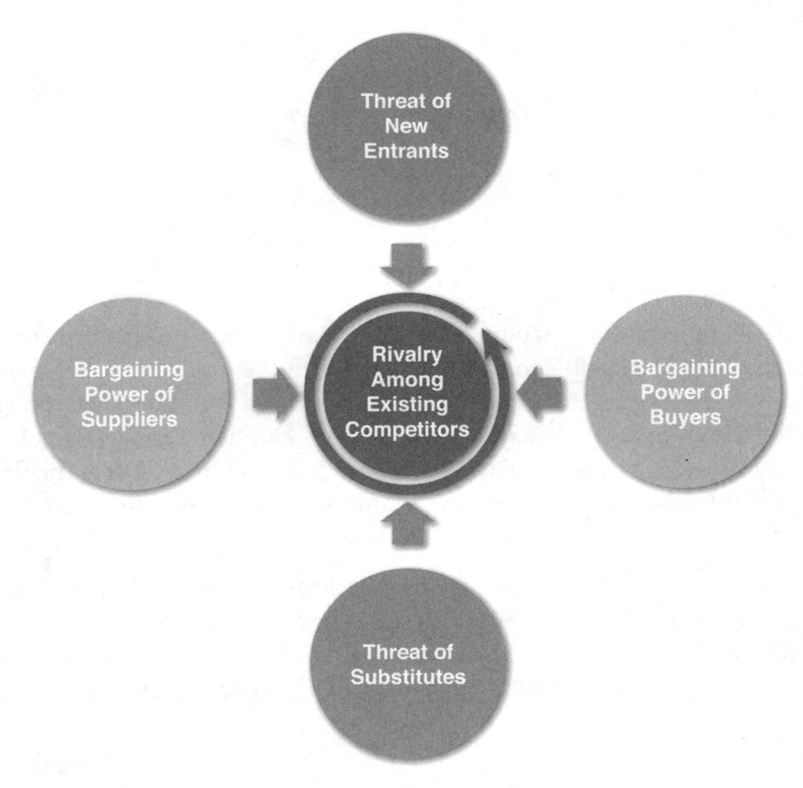

Figure 2-5 Porter's Five Forces Model[14]

Combining Porter's Five Forces and the Market Analysis

Using the spend data by supplier and by commodity provides guidance on the markets to assess. Look at each of the five forces and make a determination of where there might be problems (attractiveness and unattractiveness of each of the constructs) and where there might be opportunities.

Threat of New Entrants

The purchasing team needs to understand the barriers to entry for the market. Consider, for example, other low-cost country manufacturers entering into the field and how this impacts the incumbent manufacturers. Can these new suppliers easily enter the market, or is there a potential resource or investment hurdle that they can't overcome? Some potential barriers that would inhibit (or facilitate) the entry include availability of skilled workers, product life cycles, and risk and cost of switching. Brand equity and customer loyalty might make it challenging to enter a market; customers of Ford are an example. Government regulation, access to critical technologies, inputs, or distribution are also things to think about. The fewer barriers present, the easier it is for new entrants to the market.

Bargaining Power of Suppliers

As many supply markets begin to consolidate, fewer suppliers means that a greater amount of supplier power exists in markets, and that there is higher complexity in the purchasing process. The power position of the suppliers influence cost, negotiations, and possibly entry into the market. You need to carefully consider the supplier concentration, availability of substitute inputs, and buyers switching costs for these other inputs. The idea of integration both forward from the supplier's perspective and backward from the buyer's perspective is also something that might influence a shift in power. Another is the importance of this particular industry to the suppliers. In the grand scheme of the suppliers' business, how important is your business? Is it easy for them to give up the business and still be successful? The greater the power that the suppliers have, the more challenging it is to maintain prices, gain access to needed innovation or technology, or gain access to the appropriate suppliers.

Bargaining Power of Buyers

The power of buyers increases as specifications are consolidated and specific industry standards are implemented. The power position of the buyers can also influence cost, negotiations, availability of inputs, and accessibility to inputs. How many buyers are there relative to the number of suppliers? Is there the opportunity to use multiple sources? How important is the product to the buyer? How much volume is necessary? Answers to questions like these can help you understand the competitive landscape and determine where in the spectrum of the competition the buying organization sits. Price sensitivity, switching costs, profits, and availability of substitutes are all impactful on the power balance. A market with a powerful buying presence often can overcome barriers to entry; however, the competition also intensifies.

Threat of Substitutes

The purchasing team has to understand the availability of substitutes, relative price, quality of substitutes, and any switching costs to buyers. If a substitute becomes available, the buyer must be aware of the implication and also understand the supply chain implications of switching to a potentially new material.

Rivalry Among Existing Competitors

All the previous categories have an influence on the rivalry among existing firms. The team needs to think about the concentration of competitors, industry growth rate, fixed versus variable costs, product differentiation, and exit barriers. What is the relative size of competitors and how does your organization compare? Is there a lot of diversity of competitors?

Porter's Five Forces Model provides a way to help an organization begin to sort all the data that has been collected, including the spend analysis and market information. The purchasing team has to organize the data in a way that an effective decision will be made.

Kraljic's Portfolio Analysis

Another tool that should be applied in the assessment of the market is Kraljic's portfolio analysis.[15] The importance of this matrix was mentioned in Chapter 1. However, understanding how to use this tool for market analysis is discussed next.

The matrix takes a broader view of the individual commodities and available supply base by looking at the level of spend (importance to the organization) and the complexity of the supply market (how many capable suppliers are available and how much risk is associated with buying the commodity). Purchasers use the data collected to make a determination of where a particular commodity or service sits in the matrix. Assigning the commodity to a particular quadrant in the matrix changes the tactics used in purchasing and managing the relationship and the actions used to execute those strategies.

Using the Portfolio for Decision Making

Figure 2.6 is a more in-depth depiction of the matrix developed by Kraljic in 1983. The matrix is divided into four categories: Noncritical (low importance, low supply risk), Leverage (high importance, low supply risk), Strategic (high importance, high supply risk), Bottleneck (Low importance, high supply risk). This chapter focuses specifically on the types of relationships beneficial in each type of category. Chapter 3 involves the different strategies, tactics, and actions used to better manage each classification.

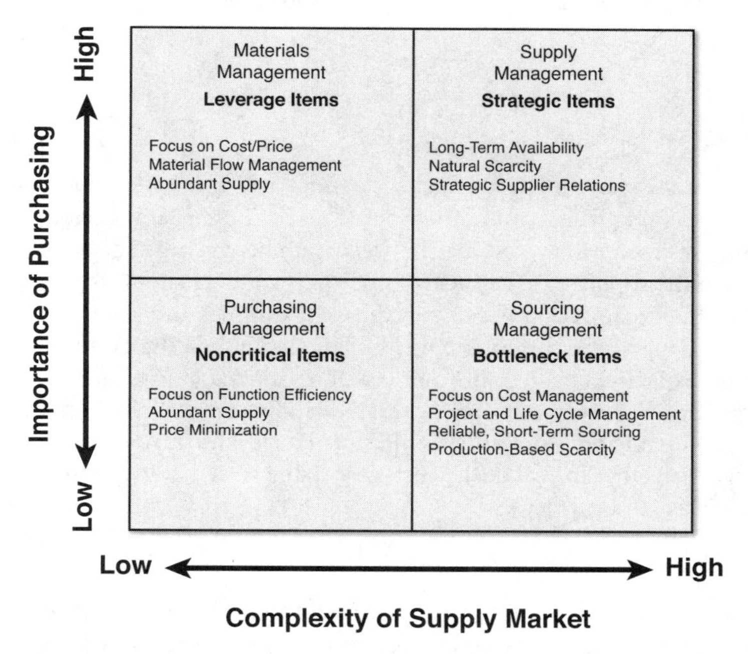

Figure 2.6 Kraljic's expanded portfolio analysis

For *noncritical*, or the more generic purchases, the focus is on finding the lowest possible purchase price from a field of many suppliers. For these types of items, there are low switching costs allowing for easy "supplier hopping." Supplier relationships are often arms-length or transactional, and contracts are short term.

Case in Point—Noncritical

Some items are routinely used in an organization but are not critical to the success of that organization. For example, office supplies would fall into this category. They are not critical to running the business. If you run out you can always replenish at a local Wal-Mart. The issue for items in this category is that you focus on the lowest price. Switching suppliers is easy and many suppliers are available.

Many companies can provide the products and services in the *leverage* category; these are the more commodity-type items. Items classified here have a high-level of spend making them important to the organization and the suppliers that provide these items into a preferred category.

Case in Point—Leverage

A lot of variety exists in the types of items classified as *leverage* items at each organization. One that is relatively common is that of packaging. Many packaging suppliers are available, so if necessary volumes between suppliers can be shifted or one can be dropped from the supply base because of poor performance. Items in this category are important to the organization because of the high volumes and level of spend. Suppliers in this category have a different type of relationship than the noncritical and try to leverage their expertise to gain more business. Logistics suppliers, packaging suppliers, and others that fall in this category generally gain a competitive edge through their relationship, which is built on trust and performance. The buyers and suppliers work collaboratively together to reduce the costs associated with making these materials with the intent to drive down prices.

Strategic or *critical items* have more complexity and risk involved in the purchase. This is often due to limited availability (that is, resource scarcity) and fewer capable suppliers. These are typically items that you need to ensure continuity of supply for your customers. Buying firms enter into long, cost-based contracts with the suppliers in these firms and may engage the suppliers early in the process of new product development. These supplier relationships are strategic.

Case in Point—Strategic

The all-wood furniture company has one commodity that it considers strategic: lumber. If the lumber is not available, the company cannot produce its products. The lumber suppliers are strategic and the relationships are long term. Because lumber is a natural resource, it is subject to natural scarcity. During times of scarcity, it is the suppliers that help the manufacturer to continue its processes. The manufacturer often talks to the supplier about inventory level, purchasing, quality standards, and so on. The supplier suggests innovative and cost-saving ideas.

Bottleneck items are typically project-oriented or unique and often require niche suppliers—for example a large piece of equipment used in mining. There is a high level of market complexity, and the management of these items is often difficult and time-consuming. The relationship is usually more transactional.

Case in Point—Bottleneck

These items are extremely time-consuming for purchasers to manage. Often they are individual projects, for example, the purchasing of equipment for a manufacturer. It takes a long time from the initial idea to delivery; few suppliers of each type of machine are available; and the logistics are challenging. IT systems can fit in this category, where the investment costs are high and the implementation is time-consuming.

Supplier Analysis

Supplier analysis is used after the portfolio analysis is complete. After this assessment, the team can make supplier recommendations. Purchasing has to identify current and potential suppliers, determine any information technology requirements, and identify opportunities to leverage the commodity expenditures with similar commodities (leverage volume and consolidate spend). The identification process and important supplier requirements can vary depending on the type of commodity, how important it is to the organization, and how much risk is associated with its purchase. The criteria may also assess capabilities to meet future goals. Many categories can be considered depending on the needs of your organization; however, some of the major categories are

- Price, cost, total cost
- Financial condition/viability/performance
- Process and design capabilities
- Management capability
- Planning and control systems
- Quality management systems
- Technical capability
- Environmental regulation compliance
- Social/ethical behavior
- Capacity

- Relationship potential
- Certifications

The criteria are both objective and subjective, and each major category has subcategories and weights (level of importance) associated with it, as discussed in Chapter 1. Each supplier is assessed or scored on the important criteria. You should use the sum of the total weights to determine potential suppliers. These criteria are generally developed in tandem with other stakeholders and weighted in terms of importance to the organization.

An example of a weighted factor rating approach is included in Figure 2-7. The weighted calculation is the score divided by scale multiplied by subweight. Sum each category and then add up those category totals. Ultimately, you can use these criteria to select the supplier and to measure and manage supplier performance.

Supplier: R&M Components					
Category	*Weight*	*Subweight*	*Score (5 point scale)*	*Weight Score*	*Total*
1. Quality Systems	20				
Process control systems		5	4	4	
Quality Certification (IS O14000)		8	4	6.4	**17.4**
Parts-per-million defect performance		7	5	7	
2. Management Quality	10				
Management/labor relations		5	4	4	**8**
Management capability		5	4	4	
3. Financial Condition	10				
Debt structure		5	3	3	**7**
Turnover ratios		5	4	4	
4. Cost Structure	12				
Cost relative to industry		4	5	4	**11.2**
Understanding of costs		4	4	3.2	
Cost control/reduction efforts		4	5	4	
5. Delivery Performance	12				
Performance to promise		4	3	2.4	**7.2**
Lead-time requirements		4	3	2.4	
Responsiveness		4	3	2.4	
6. Technical/Process Capability	18				
Product innovation		5	4	4	**15.4**
Process innovation		5	5	5	
Research and development		4	5	4	
Technological Innovation		4	3	2.4	
7. Information on Systems Capability	12				
EDI capability		4	5	4	**8.8**
CAD/CAM		2	0	0	
e sourcing capabilities		6	4	4.8	
8. General	6				
Support of minority suppliers		2	3	1.2	**4.8**
Environmental compliance		2	5	2	
Supplier's supply base management		2	4	1.6	
				Total Weighted	**79.8**

Figure 2-7 Supplier scorecard, R&M components[16]

Supplier Evaluation and Selection

Most purchasing experts agree that there is no one best way to evaluate and select suppliers, and organizations use a variety of different approaches. Regardless of the approach employed, the overall objective of the evaluation process should be to reduce purchase risk and maximize overall value to the purchaser. Most corporations have a standardized supplier evaluation and selection process; the verbiage and number of steps may be different, but the activities within each step are similar.

1. **Recognize the need for supplier selection.** Many reasons exist for an evaluation and selection decision. For example, new product development requires a new pool of suppliers. You need to consider many issues such as current supplier capacity or performance. Building new facilities or moving facilities to a different location can also require new suppliers.
2. **Develop a cross-functional sourcing team.** After a need is identified, the people with an interest in a successful sourcing event must be identified and begin work as a cross-functional team. This team works closely together in identifying, selecting. and evaluating suppliers. Typically, members of cross-functional teams are evaluated on joint metrics (which are discussed in Chapter 6).
3. **Identify key sourcing requirements.** These are the major "knock-out" criteria for initial screening. For example, a supplier might be too small to meet your requirements. Perhaps they don't have the necessary capabilities. Potentially they are struggling financially and are too much of a risk. The sourcing team determines what is important, and these criteria are used during the initial screening stage.
4. **Determine sourcing strategy.** This has to align with both corporate strategy and fit within the parameters of the other work already completed during earlier assessments. For example, is this a short-term or long-term relationship? Will the supplier be domestic, nearshore, or offshore? Will this be a sole source situation, or is that too risky of a proposition?
5. **Identify potential sources for supply.** Depending on the classification of the commodity, more or less time might be involved at this step. The more strategic the item and the lower the capabilities of the current suppliers, the more time is spent here. The team is responsible for gathering a pool of suppliers that can provide the good or service within the specified parameters.
6. **Limit suppliers in the selection pool.** This step occurs as the information is gathered and the critical issues are assessed. The purchaser may have many potential sources from which to choose but the performance capabilities of suppliers vary. As discussed in earlier sections, the key here is to understand the subjective and

objective criteria that are most important for the commodity or service purchased. An in-depth financial risk analysis of each potential supplier is also necessary. When the pool is at a point that it is manageable, supplier site visits might help to eliminate some of the remaining contenders.

Following are potential financial metrics/measures for risk assessment and supplier evaluation:

- Credit rating
- Average collection period in days
- Stock price
- Average accounts payable in days
- Return on Investment (ROI)
- Inventory turnover
- Cash flow
- Average days in inventory
- Working capital
- Fixed asset turnover
- Current and quick ratio
- Return on assets
- Profit and operating profit margin
- Return on equity
- Debt to equity
- Times interest earned
- Cash to cash cycle
- Year over year change in costs versus change in revenue
- Gross margin ROI

7. **Determine the method of supplier evaluation and selection.** At this stage, the pool of available suppliers is more manageable. The team has to decide how to evaluate this smaller pool of suppliers. A well-structured site visit is highly recommended to each of the remaining capable suppliers. However, the decision to visit all suppliers requires a commitment of resources that are sometimes not available.

8. **Select supplier and reach agreement.** After all the assessment, much discussion among group members, and an analysis of supplier "fit" with the buying organization, the supplier is selected. It is time to work on the negotiation and ensure that all the specifications are accurate.

Contract Negotiation and Management

Negotiation is an exploratory and bargaining process (planning, reviewing, analyzing, and compromising) involving a buyer and seller, each with her own viewpoints and objectives. The goal of negotiation is to reach a mutually satisfactory agreement on all phases of a procurement transaction—including (but not limited to) price, service, specifications, technical and quality requirement, freight, and payment terms. Not all commodities or services require negotiations, so the decision to negotiate relies on the effective classification of the goods and service in the commodity matrix.

The basic steps in the negotiation process are listed next.[17] As with many purchasing activities, it is a cross-functional team effort. Other team members might include cost accountants, engineers, business analysts, and others.

1. **Select the team and assign the chief negotiator.** The team is there for backup, often performs data analysis, and helps the primary negotiator make decisions.
2. **Determine objectives, including a win-win outcome for both parties.** A win-win is important if the relationship with the supplier is key, for example, if this supplier will receive a high volume of business, or the supplier has a key technology available. If suppliers feel like the situation is not a win for both, they either will not accept your business, or you may have a potential trade-off with quality.
3. **Prepare for the negotiation by looking at comparative bids, potentially going on a site visit, and assessing industry price trends.** (Information is power, so the more you can find out, the better.) During this stage the purchaser should develop a proposal, think about potential questions from the supplier, and gather any information on past purchases with the supplier. Establishing an objective target for price and putting boundaries on it can help to control the negotiation.
4. **Determine bargaining strength, and think about the bargaining strength of the supplier.** Go back to the data collected during the market assessment to see if you can understand who is in the position of power.
5. **Plan the agenda, the place, the time, and the minimum and maximum positions.** The better the plan and the more information available can help establish a position of authority.

6. **Set a negotiation strategy based on a list of obtainable objectives or goals.** The choice of the strategy will depend on what is at stake, the history with the other party, and the actual objectives being pursued.
7. **Determine particular tactics on how to achieve the strategy.** Understand that this is not a game; there should be no dishonesty and no psychological tricks. Purchasers have to act ethically in all aspects including negotiations; be honest and don't reveal competitors prices. Never guess about volume or overstate potential volume. Many purchasers have found themselves in situations in which they allude to high volumes, or reduced quality expectations, and then when it comes time, this doesn't happen.
8. **Perform a post-negotiation debrief, develop an action plan based on the agreement, and assess the performance of the negotiating team.** Think about how to achieve continuous improvement in the negotiation process. In addition, understand that not everyone is a good negotiator, and no matter how much training you have won't guarantee that you achieve chief negotiator status. However, the skills in data analytics and preparation are needed to achieve a successful negotiation.

You want the supplier to win for a long-term, mutually rewarding relationship. If suppliers feel "beaten down" during a negotiation, it may be difficult for them to meet the exacting specifications.

Special consideration should be given to negotiations with companies from other offshore locations, which is discussed further in Chapter 5. Language and cultural nuances can often impede or even disrupt a negotiation. Companies spend time and money training people to negotiate in a number of different locations. Part of your planning process should include a country analysis and discussion with others that have negotiated in different areas of the world.

Managing the contract requires intensive communication and collaborative buyer-supplier relationships. Suppliers are responsible for timely and satisfactory performance of their contracts. Unfortunately, the purchaser can't rely entirely on the supplier to ensure that work is progressing as scheduled and that delivery will be as specified. Poor performance or late deliveries disrupt production operations and can result in lost sales. Supply management must monitor supplier progress to ensure that deliveries are met. The level of monitoring depends on the type of commodity, the criticality of the material or service, and the capability of the supplier to meet the buying firm's requirements.

You can effectively monitor supplier performance in a number of ways, such as face-to-face meetings, operations progress reports provided by the supplier, Gantt charts (Figure 2-8), PERT charts (Figure 2-9), quality audits, and others. Contract management can be a challenge, especially if you have large inventories to cover up quality and delivery problems. Purchasers can't make assumptions that everything is going as planned.

Gantt Chart - Project Development

Activity Name	1st Quarter			2nd Quarter			3rd Quarter			4th Quarter		
	Jan	Feb	Mar	Apr	May	June	July	Aug	Sept	Oct	Nov	Dec
1. Design and Development												
2. Cost Estimates												
3. Construction												
4. Testing												
5. Celebrating												

Figure 2.8 Gantt chart for product development[18]

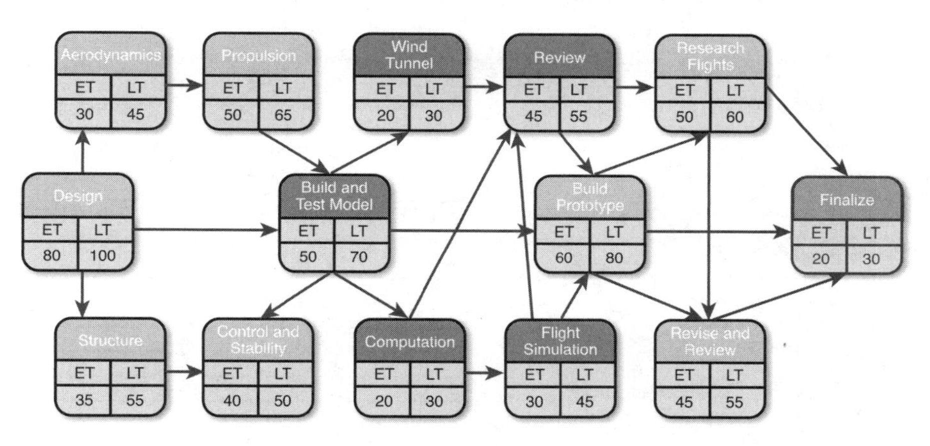

Figure 2.9 Pert chart for an airplane design process[19]

Supplier Relationship Management and Development

As discussed in earlier sections, every organization has a broad spectrum of supplier relationships that range from transactional or arm's length through collaborative to strategic alliances. Developing and managing relationships can be both challenging and fulfilling.

One key to relationship management is a structured supplier performance evaluation.[20] The purpose of these evaluations is to both manage performance and enhance the relationship with the supplier. Typically, organizations evaluate performance on a 3-to-6-month moving average. This is so suppliers' mistakes don't haunt them forever. There are four types of evaluation plans that are common: categorical, weighted-point, cost ratio, and a total-cost based approach.[21] Illustrative examples of each are discussed in the following sections.

Categorical Plan

Figure 2-10 shows an example of a categorical plan supplier evaluation. The categorical plan relies heavily on the judgment and experience of the decision makers involved in the supplier relationship. The supplier performance is reviewed periodically by an evaluation committee composed of all representatives.

Depending upon the performance, the supplier is given a plus point, neutral mark, or minus sign. The performance trends over a period of time are built up, and the supplier with the increasing trend of a plus point is chosen. A preponderance of pluses or minuses needs notification to supplier with comments.

On the basis of experience and periodical meetings, you can establish a list of factors on the suppliers' performance in each area and give each factor a grading as never, seldom, usual, always, and so on. This system, though nonquantitative, provides a means of systematic record keeping on performance criteria.[22]

Supplier:	**Date:**		
Summary Evaluations, by Department	**Preferred**	**Neutral**	**Unsatisfactory**
Supply Management			
Receiving			
Accounting			
Engineering			
Quality			
Performance Factors			
Supply Management			
Delivers on schedule			
Delivers at quoted prices			
Has competitive prices			
Anticipates our needs			
Helps in emergencies			
Does not request special consideration			
Currently supplies prices, catalog and technical information			
Advises us of potential troubles			
Has good labor relations			
Keeps promises			
Receiving			
Delivers per routing instructions			
Has adequate delivery service			
Has good packaging			
Accounting			
Invoices correctly			
Issues credit memos punctually			
Engineering			
Has past record on reliability of products			
Has technical ability for difficult work			
Provides quick and effective action in emergencies			
Furnishes requested data promptly			
Quality			
Provides high-quality material			
Replies with corrective action			

Figure 2-10 Example of a categorical plan supplier evaluation

Weighted Point Supplier Evaluation

The weighted point supplier evaluation, similar to the weighted scorecard discussed earlier, uses predetermined factors and weights (or how important those factors are to the evaluation), and calculates and compares that to actual performance of the supplier. Figure 2-11 shows an example of a weighted point supplier evaluation.[23]

R&M Monthly Performance Evaluation			
Factor	Weight	Actual Performance	Performance Evaluation
Quality	50	5% rejects	50 X (100 – .05) = 47.50 (a)
Service	25	3 failures	50 X [100 – (.07 X 3)] = 19.75 (b)
Price	25	$100	50 X ($90/$100) = 22.50 (c)
			Overall Evaluation = 89.75

Figure 2-11 Example of a weighted point supplier evaluation[24]

Total Cost Evaluation

Using a total cost model to initially assess the cost of doing a business with a supplier is important. However, a purchaser can evaluate the supplier by comparing actual with expected costs. The purchaser can also compare the supplier for their involvement in cost reduction efforts. Figure 2-12 is a brief example of a total cost model.

	Cost	Unit of Measure	Total Cost
Price paid for each boom box	$42	Each	$33,600
Cubes per boom box	3	Each	800
Boom Boxes per pallet	80	Pallet	10
Shopping cost to Los Angeles Port	$4750	Container	$4750
Cubes per container	2400	Container	2400
Customs fees 3% the value of the container	3%	Container Value	$1008
Duties 5% of the container	5%	Container Value	$1680
U.S. Port Handling per container	$1750	Container	$1750
Insurance Per Container	$600	Container	$600
Transportation port to Warehouse	$1200	Container	$600
Warehouse movement and Storage Cost	$1200	Pallet	$1200
Quality Control	$10	Each	$8000

Figure 2-12 Example of a total cost evaluation for boom boxes

At the same time that firms are outsourcing more materials, sub-assemblies, finished goods, and services to other organizations, they are increasing their expectations for suppliers to deliver innovative and quality products on time and at a competitive cost. When a supplier can meet these needs, three alternatives are available: in-source; re-source and find a new supplier; and help improve the existing supplier's capabilities. Depending on the type of commodity, the best option might be to switch suppliers (if switching costs are low). However, for the more collaborative and strategic relationships helping a supplier to improve (or developing the suppliers capabilities) is the best option. At this point in the purchasing process, significant investment has already been made in the supplier relationship so that it makes sense to work collaboratively on change.

Supplier development is any activity that a buying firm undertakes to improve a supplier's performance and capabilities to meet the buying firm's supply needs. There are a number of tools you can use and a number of activities that you can do to help in these efforts, including the following:

- Assess supplier's operations.
- Provide incentives to improve performance.
- Instigate competition among suppliers.
- Work directly with supplier through training and other activities.
- Co-locate buying firm employees at the supplying firm.
- Involve suppliers in new product and process development at the buying firm.
- Provide improvement-focused seminars for suppliers.
- Provide tooling and technical assistance for suppliers.
- Share the savings from development improvements.

In some situations supplier development is a catalyst of process change for buying organizations.

There are many barriers to effective supplier development.[25] In many cases, suppliers don't want to be developed or don't feel that they need development. Many of the barriers revolve around poor communication and feedback. There is also a feeling of "blaming the supplier for the problems." To avoid problems with development, ensure that there are clear expectations, metrics aligned between sourcing and performance, and clear and frequent communication.

Supplier Performance Measurement and Evaluation

There are hundreds of purchasing and supply chain measures. Some of the more common categories of measures are in Table 2-2, which is a sampling of measures to evaluate supplier performance. Keep in mind that some measures are more or less appropriate for some suppliers. Chapter 6 defines the measures in more depth and discusses the implementation of the measures.

Table 2.2 Supplier Performance Measures

Category for Measurement	Sample Measures
Price performance	■ Actual price compared to plan ■ Actual prices versus market index
Cost effectiveness	■ Cost changes ■ Cost avoidance
Revenue	■ Supplier contribution as a reason for new business
Quality	■ Parts per million (PPM) ■ Field failure rates by supplier
Time/delivery/ responsiveness	■ On-time delivery and responsiveness ■ Responsiveness to schedule, mix, and design changes
Technology or innovation	■ First insight of new technology ■ Standardization to reduce complexity
Physical environment and safety	■ Sustainability criteria in procurement decisions ■ Maintaining appropriate records to feed into corporate sustainability and social responsibility reporting
Asset and integrated supply chain management	■ Inventory turnover ■ Transportation cost reduction

Many firms have made great progress in supplier evaluation and measuring performance. Purchasers can track quality, cost, delivery, and other performance areas. Also, firms can now quantify nonperformance. Systems are becoming more sophisticated, and purchasers can better measure the total cost of doing business with a particular supplier.

Conclusion and Chapter Wrap-Up

This goal of this chapter is to bring together the key processes involved in managing the activities and processes involved in supply management. This includes their involvement in managing risk and market, category, and portfolio analysis. Many new tools are introduced such as Porter's Five Forces Model and spend analysis. Also, Kraljic's matrix is again used to show how purchasers can better manage the supplier relationships.

The role of purchasing in an organization is becoming more strategic and more critical for success. The selection, evaluation, and management process is time-consuming and resource-intensive. However, the costs of choosing the wrong supplier can be catastrophic to an organization and can open up the organization to financial, reputational, and operational risk. After suppliers are selected and added to the portfolio of available suppliers, purchasers need to make sure that they get what they contract for. Remember that the actions of your suppliers are a reflection on your performance. Many of the ideas introduced in this chapter are discussed more in-depth in the following chapters. Some of the key points from this chapter include the following:

- Understanding the basic processes involved in managing supply operations
- Determining how to establish a commodity strategy and perform a category analysis
- Performing a supplier screening and selection
- Understanding how to negotiate and manage contracts
- Validating supplier performance and quality
- Developing and managing supplier relationships appropriately
- Measuring supplier performance and managing supplier development

Key Terms

- Operational risk
- Financial risk
- Reputational risk
- Supply risks
- Commodity strategy development
- Procure to pay process
- Spend analysis

- Porter's Five Forces Model
- Kraljic's 1983 portfolio analysis
- Supplier evaluation scorecard
- Noncritical
- Leverage
- Strategic
- Bottleneck
- Supplier analysis
- Portfolio analysis
- Negotiation
- Structured supplier performance evaluation

References

1. Monczka, et al. (2011). *Purchasing and Supply Management*, 5th Edition. Southwest Publishing: Mason, OH.
2. Ernst & Young (2010). "Five Areas of Highly Charged Risk for Supply Chain Operations." Retrieved September 12, 2013, from http://www.ey.com/Publication/vwLUAssets/Five_areas_of_highly_charged_risk_for-supply-chain-operations/$FILE/Climate%20change%20and%20sustinability_Five%20areas%20of%20highly%20charged%20risk%20for%20supply%20chain%20operations.pdf.
3. Campbell, A. (2013). "Top 10 Organizational Risks for 2013," Risk.net. Retrieved on September 6, 2013, from http://www.risk.net/operational-risk-and-regulation/news/2228959/top-10-operational-risks-for-2013.
4. Information was presented by a representative from Caterpillar, September 11, 2013, in Knoxville, TN.
5. Information was presented by representatives from Rolls Royce, September 4, 2013, in Knoxville, TN.
6. Financial Management (2011). "8 Ways to Reduce Supply Chain Risk." Retrieved September 6, 2013, from http://www.fm-magazine.com/feature/list/8-ways-reduce-supply-chain-risk.
7. Caterpillar, (2013), ibid.
8. Monczka, et al. (2011), ibid.

9. PWC Oracle Practice. "Optimize Procure to Pay Processes for Profitability, Efficiency, and Compliance." Retrieved August 31, 2013, from http://www.oracle.com/us/products/applications/ebusiness/optimize-procure-to-pay-processes-1855140.pdf.

10. Procure to pay process flow figure retrieved on August 31, 2013, from Google images for the procure to pay process.

11. Monczka, et al. (2011). *Purchasing and Supply Management*, 5th edition. Southwestern Publishing: Mason, OH.

12. Burt, Dobler, Starling (2003). *World Class Supply Management*. McGraw Hill: New York.

13. Porter, M.E. (1980). "Industry Structure and Competitive Strategy: Keys to Profitability," *Financial Analysts Journal*, pp. 30-41.

14. Porter, M.E. (1980). "Competitive Strategy."

15. Kraljic, P. (1983). "Purchasing Must Become Supply Management," *Harvard Business Review*, September-October, pp. 109–117.

16. Adapted from Monczka, et al (2011).

17. Burt, Dobler, Starling (2003). *World Class Supply Management*. McGraw Hill: New York.

18. RFFlow5, Personal Flowcharting, Gannt Chart Product Development. Retrieved on September 12, 2013, from http://www.rff.com/project_development.htm.

19. SoftIndeed, Airplane Design Process, Pert Chart. Retrieved on September 12, 2013, from http://www.softindeed.com/shop18/components/com_virtuemart/shop_image/product/PERT_Chart_Exper_4f9e934384864.gif.

20. Wheaton, G. (2009). Supplier Performance Management. Retrieved on September 13, 2013, from http://www.epiqtech.com/supplier-performance-management.htm

21. Monczka, et al. (2010), ibid.

22. Ghosh, P. (2013). "8 Major Rating Plans are Used for Vendor Rating." Retrieved on September 24, 2013, from http://www.shareyouressays.com/116381/8-major-rating-plans-are-utilized-for-vendor-rating.

23. Monczka, et al (2011), ibid.

24. Burt, Dobler and Starling (2005). Ibid. (a): 100%: Percentage of rejects; : 100%: 7 percent for each failure; (c): Lowest price offered or price actually paid.

26. Handfield, Krause, Scannell, and Monczka. "Avoid the Pitfalls."

3

PRINCIPLES AND STRATEGIES FOR ESTABLISHING EFFICIENT, EFFECTIVE, AND SUSTAINABLE SUPPLY MANAGEMENT OPERATIONS

This chapter describes the procurement principles and strategies that help properly align an organization's supply management operations with its supply chain strategies. The key components of a sourcing strategy, the role of supply management in the outsourcing decision, and sustainability are introduced.

Learning Objectives

After completing this chapter, you should be able to:

- Understand supply management principles and strategies.
- Describe and define the key components of a sourcing strategy for each of the procurement categories and how these relate to organizational strategies.
- Determine how and when to rationalize and optimize the supply base.
- Determine the role of supply management in the insource (make) versus outsource (buy) decision. Describe the risks and benefits of each.
- Prepare and perform the appropriate cost management analyses.

- Learn about other analytic tools to manage the supply chain.
- Examine the key components of sustainability and how supply management decisions impact the environmental footprint of the organization.

Introduction to Supply Management Principles and Strategies

The process of aligning supply management goals with corporate objectives is especially important for supply management and supply chain managers. To be successful, the strategy development process has to be integrative. The strategy should be drafted by and have significant input from those people responsible for implementation.[1] The corporate level strategies must flow into functional level strategies. For supply management, these corporate levels strategies also have to be tied in to goals for particular commodities or services.

It is imperative that strategic sourcing takes its direction for building sourcing strategy from business unit requirements; these are the internal customers. Supply management must add value to the customer (both internal and external) in terms of implementing specific objectives to achieve the goals of the business unit.

Category Sourcing Strategies

As discussed in Chapter 1 and Chapter 2, different classifications of commodities and services have different types of sourcing strategies and objectives. Generally, these strategies and objectives are developed using a cross-functional team. This means that those who are involved are familiar with or have a vested interest in the commodity or service being purchased. The cross-functional team has to operate in an effective way so that the value of the participant's time maximizes corporate objectives. Some key activities are involved in establishing the appropriate and therefore more effective cross-functional team to include the following:

- Tailor membership on the cross-functional team to the assigned activity.
- Make sure that the team members have the right skills, experience, and training.
- Guide team efforts with performance measures and goals.
- Provide needed organizational resources.
- Grant authority to control internal activities and make substantive decisions.
- Appoint a formal team leader.
- Link team members' performance evaluations and compensations to team performance.[2]

Chapter 2 discusses different methods for educating team members about supply market conditions, forecasted spend, and the user and stakeholder requirements. After these steps are completed, a portfolio analysis is used to better structure and segment the supply base. The portfolio analysis classifies suppliers into one of four types: strategic, transactional, preferred, and arms-length. These align with the appropriate type of supplier relationship.

This segmentation or classification into the matrix enables the purchasing team to better understand how important the item is to the business and therefore propose an appropriate strategy for managing the commodity and the associated supplier relationships. Each strategy then determines the tactics and the actions most appropriate for that particular quadrant.[3] Hence, each quadrant has a strategy, particular tactics to support that strategy, and actions to implement the strategy. Kraljic's portfolio matrix is again adapted and re-introduced in Figure 3-1. The strategies, actions, and tactics are discussed by quadrant in the paragraphs that follow.

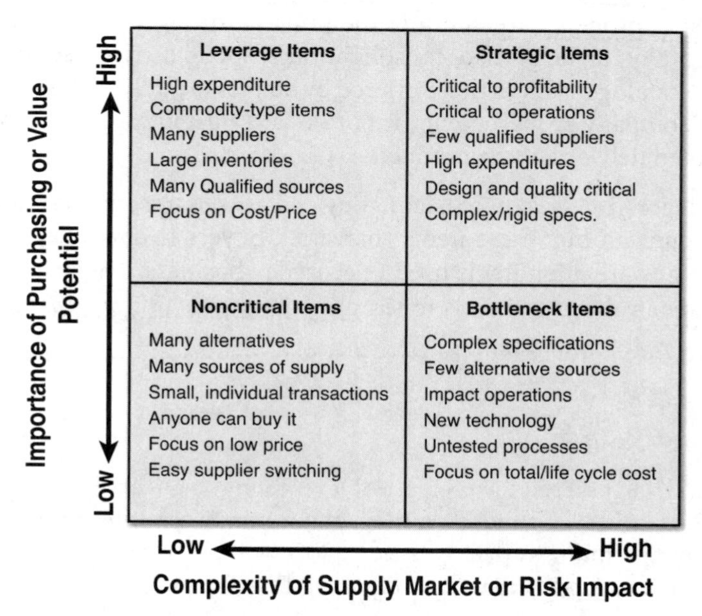

Figure 3-1 Kraljic's expanded matrix[4]

Using Kraljic's Matrix to Implement Strategy

This section explains the relationship between categorization of the commodities into a specific quadrant and effective implementation of strategies within those quadrants. Keep in mind that the classification within a quadrant is company-specific; this is not a one-size-fits-all strategy. For example, packaging might be a leverage item for most manufacturers. However, for FedEx, the right type of packaging, the right colors, the right packaging weight, and other specifications might make it a strategic item.

Noncritical—Arm's Length Supplier

Products and services classified in this category are generally readily available and often are relatively low in importance to the organization. As mentioned in previous chapters, office supplies can easily be classified here for most organizations. For a company such as Dell or Apple, the power cords can potentially be classified in this category.

There are many suppliers available on the market, and the costs to switch suppliers are relatively low, causing low purchase price to be the driver of the purchasing decisions. Purchasers are generally measured on purchase price compared to market price (performance measurement is covered in later chapters) or purchase price variance.

The best tactic to support items in this category is to reduce the amount of time spent on the purchasing process. Streamlining the process through automation or supplier-provided technology can help. Also, consolidation of spend can help support a low-price strategy. Online catalogs, such as those offered by Staples, Best Buy, or Office Max, enable buyers to use a company procurement card or P-Card (just like a credit card), often with free and expedited delivery of the products.

Items in this category rarely require negotiating and are often managed under short-term contracts. Anyone can buy these items (maverick buyers in this category often cause problems) and they are often items used in everyday business. The primary purchasing strategy for items in this category is to simplify the acquisition process and reduce the buying effort.

Case in Point—Noncritical

Many companies have faced the problem of "everybody's a buyer." Employees purchase something for business use, and then they ask for reimbursement through petty cash. There are a lot of administrative costs associated with this process, such as setting up accounts, reconciling petty cash, and ensuring that the purchases are legitimately a business purchase.

After the P-card was introduced, it eliminated many of the issues associated with maverick buying. Purchases became more centralized, and reconciliation of accounts was much easier. P-cards also eliminated the more personalized buying, such as a particular type of pen for a particular person, and it enabled consolidation of spend, and therefore, lower prices. Items in this category should be purchased using a streamlined process and a focus on price reduction. Many organizations mandate a pre-approval process for any individual purchase not using a P-card.

Leverage Commodity—Preferred Supplier

Because volume (spend) is high for this category, it is important to the buying organization (and there is opportunity for significant savings) if the suppliers are managed appropriately. As mentioned previously, packaging for most organizations falls into this category. For Dell or Apple, potentially the keyboards, mouse devices, or screens fall into the leverage category.

Because these items are important to the organization primarily because of the amount of money spent, quality and cost management are key considerations. The primary strategy for purchasers for items in this category is to maximize opportunities to reduce costs. This requires leveraging both volume and spend, or concentrating the business with fewer suppliers, thereby introducing a more competitive environment.

Purchasers can take many actions to implement the appropriate strategy including competitive bidding and reverse auctions. (The business goes to the lowest bidder that meets the defined specifications.) Because many suppliers in the market provide these leverage items, the competitiveness keeps prices down. The increased volumes in this category make the buying company a preferred customer to the supplier. The relationship is far more collaborative here than with noncritical items, and both buyer and the supplier work together to find ways to develop cost-savings and improve processes.

Case in Point—Leverage

A company located in the Western United States was discussing suppliers of call center services. Each of six different business units was represented in the discussion. Each of the business units had a person responsible for selecting and managing the call center supplier. These people did not work in the purchasing area but worked for the business unit.

As the managers started to discuss the purchasing spend in this area (it was significant), they realized that some of the business units were using the same supplier but paying a higher price per minute. When the supplier was questioned about this practice, its response was that some of the business units were easier to work with, and this is why the lower price. The company realized that it had a problem and called in the supply management department to help.

Supply management decided to begin the strategic sourcing process for call center services and put out an RFP, wondering who wanted the business. Ultimately, a supplier was selected that could support all six business units. The spend was directed to a single supplier with call centers in various regions. The final result was an optimized supply base with more leverage and a lower price.

Strategic Commodity—Strategic Supplier

The idea here is to develop the supplier's capabilities and increase the role of selected suppliers into the process of continuous improvement and potentially new product development. For every company, the strategic items are classified differently. These are often the items that, if they were not available, would make it so the company could not produce the items it needed to produce. The lumber for the furniture company was mentioned earlier. For Dell or Apple, it is potentially the microprocessors.

To effectively manage these types of items, strong analytical skills and relational skills are needed. This is an important supplier, and the purchaser must minimize the organization's exposure to risk, especially in terms of disruptions.

Managing the relationship in a collaborative way helps the buying organization to reduce costs, improve products and services, and increase innovation. These relationships are managed using long-term contracts. Suppliers in this category are often involved in new product development and design, even in the early stages. Also, purchasers look to these suppliers for innovative ideas and continuous improvement.

Case in Point—Strategic

The furniture company discussed in earlier sections made solid, pine wood furniture and needed a stable source of supply for the lumber that went into making the finished products. Lumber is a naturally scarce material, and there are few suppliers of the lumber from sustainable forests. They have to be managed appropriately and collaboratively. Without the specific quality and grade of lumber, the furniture manufacturer could not manufacture the furniture. Because of this, the company formed strategic relationships with the lumber company, trying to ensure a continuous source of supply. The company also worked with the lumber suppliers to form contingency plans, in case of regulatory changes or supply problems. These suppliers were also involved in design and made a number of suggestions for changes to improve both product stability and reduce waste and cost.

Another strategic item is the finish (or color) that goes on top of the furniture. The products must be consistent in color and must match across the product line. The company tried to focus its effort on being a good citizen and minimizing GHG emissions. These suppliers suggested a way to maintain the color of the product but use a finish that was water-based. The water-based finish required new equipment and techniques to finish the products; however, the waste (or loss) of the organically based stain was significantly reduced. The company collaborated with the finish suppliers for both the domestic production and the production in China, giving most of its finishing volume to a sole source supplier that had strong, strategic ties to the organization.

Bottleneck Commodity—Transactional Supplier

As mentioned in Chapter 2, bottleneck commodities are difficult and time-consuming to manage. These items might be project-based, they might be a one-time purchase, or they might be a specialty item that has little to no commonality across the product line. These relationships are often transactional with a short-term or a one-time contract. The supply manager should try to figure out ways to shift the products out of this quadrant into one that is easier to manage. For example, consolidating spend here with a supplier would push it up into the categories on the top of the matrix. Or potentially developing and locating additional suppliers capable of providing the products might shift it over into a different category on the matrix.

Case in Point—Bottleneck Supplier

A company needed a number of different types of environmental projects completed to ensure compliance with regulations and also help meet its sustainability goals posted on the corporate website. Each of the projects was different and required various types of knowledge. Different people, in different functions within the organization, managed each project individually. Purchasing was not even involved in the initial project management.

One supplier of these environmental projects realized that the expertise of his organization was not effectively utilized and talked to his representative at the buying firm. As new projects started to develop, instead of searching for yet another supplier of environmental project services, the incumbent supplier was approached to see if the company was interested and capable. Ultimately, most of the environmental projects were funneled to this company, increasing the importance of the spend and changing the relationship with the supplier to one that was more collaborative and strategic.

Supply Base Rationalization and Optimization

Organizations should consider the optimal number of suppliers that they need to maintain to continue to do business around the globe. Supply base rationalization is the conscious process of identifying how many and which suppliers a buyer will maintain.[5] Supply base optimization involves an analysis of the supply base to ensure that only the most capable and highest performing suppliers are kept in the supply base after it is rationalized.[6] Supply base rationalization and optimization involve eliminating those suppliers that are unwilling to achieve (or incapable of achieving) specific performance objectives. The process of rationalization and optimization is ongoing.

A rationalized and optimized supply base has a number of advantages. Most important, this time-consuming and challenging process should result in improvements in cost, quality, delivery, and information sharing between buyer and supplier. With improved relationships between buyer and supplier, contracts can be extended, and also increased joint improvement efforts and a transfer of innovation and knowledge can occur. Table 3-1 contains the advantages and potential risks of a rationalized, optimized supply base.

Table 3-1 Advantages and Potential Risks of a Rationalized, Optimized Supply Base[7, 8]

Advantages	Potential Risk
Economies of scale	Supply disruption
Lower total cost	Capacity and capability constraints
Lower administrative costs	Reduced competition
Decreased supply risk	Supplier dependency
Use of the best suppliers	Quality
Full-service suppliers	Reputation

Case in Point—Supply Base Optimization

Organizations trying to harness the power of supply chain first look at internal data and performance history of the supply base. It is often obvious that the supply base is not the right size for effective management or uninterrupted supply. Redundancy often exists in certain commodities and single sources of supply in other areas. New suppliers are added without valid justification when existing suppliers had the capability and capacity to meet requirements. The analysis alone helps to create an understanding of the need for optimization and the benefits of optimization.

An optimized supply base has helped companies to increase bottom-line performance through improved cost, quality, and scheduling. One company estimated more than $50 million dollars in savings from this supply management strategy. After the spend analysis is done, a good next step is benchmarking to better understand who is best-in-class and what the best-in-class are doing in terms of optimizing the supply base.

Raytheon went through this process and developed a current and expected future state of supply base optimization and the strategies needed to move from one stage to the next. Table 3-2 shows the basic strategies taken to move from a clerical supply management function to a cutting-edge, strategic decision-maker.[9]

Table 3-2 **The Progression to a Cutting Edge Optimize Supply Base**

				Cutting Edge
Supplier Strategy			**Proactive**	Fully integrated supply strategy-early supplier involvement in design
Supplier Management			Program vs. global supplier management	Global supplier management
Social and Environmental Responsibilities		**Reactive**	Meets environmental and social responsibilities	Social and environmental responsibility incorporated into commodity strategies
SCM Skill Set		SCM lacks technical product expertise	Limited SCM product expertise	SCM actively recruits technical experts and understands commodity/industry trends
SCM Involvement in Supplier Selection	**Clerical**	SCM not involved in source selection	SCM involvement in key supplier selection	Cross-functional trams engaged in commodity strategies/supplier selection
Process	No process defined	No make/buy process defined	Make/buy strategy defined	Supply base optimized and aligned with business strategy
Bottom-Line Impact	Sub-optimized bottom line	Award of negative impact. but not quantifiable	Favorable, but not optimized impact	Demonstrated positive impact to bottom line directly associated with optimization efforts
Relationships	Transactional	Directed procurements	Preferred supplier lists	Collaborative and alliance
Emphasis	Convenience procurement	Parts availability and purchase price	Cost, quality, and timeliness	Total cost of ownership
Data	Is not available	Retrieval difficult and manual	Facilitates sourcing and pricing	Facilitates strategic sourcing/planning

Insource Versus Outsource (Make Versus Buy)

Insourcing versus outsourcing is another strategic decision that an organization makes. Adding to the decision is the complexity of the specific geography of where to outsource or offshore. The impact of this decision has long-term implications to an organization;

hence the wrong decision can have disastrous results. At an increasing rate, the insource/outsource decision-making process is liberally applied throughout organizations to all products, components, subassemblies, and services. Given that there is so much confusion in terminology used when discussing this phenomenon, following are some of the more commonly used terms in this area:

- **Insource**—To produce a component, assembly, or service internally
- **Outsource**—To purchase a component, assembly, process, or service from an outside supplier
- **Offshore**—To purchase or relocate manufacturing to a different geographic location to take advantage of lower costs
- **Reshore**—To return manufacturing or purchasing to the home country
- **Nearshore**—To return manufacturing or purchasing to a country geographically close to the home country (such as Mexico or Canada for the United States)

Like many other decisions in supply management, this one is performed with a highly effective cross-functional team. Purchasing's role is to provide market and supplier analysis and help the team make an informed decision. Purchasing is also a key player in managing the business relationship with the supplier. Careful consideration of the right supplier is critical.

One primary risk associated with outsourcing is suppliers who might try to exploit the leverage they gain through the relationship. There are two other factors to consider, one of which is core competency. The question becomes: Is this something that should be outsourced? The other issue involves making the product or service too generic, which might result in relying on your suppliers to manage the technology and delivery effectively.

Case in Point—Outsourcing Risk

A call center supplier located in India was surprised when a buying organization handed it a 200-page contract to review to enter into a contract. The supplier spent a few days trying to understand all the terms and conditions associated with the lengthy contract. One thing the supplier noticed was that "the call center operator with the longest idle time will receive the next call." The buyer had technology that was linked in to numerous supplier locations all around the world.

The buyer's strategy was to provide outstanding customer service, 24 hours per day, 7 days a week. The new supplier in India decided that it would try to help the buyer meet its goals by increasing its base of agents. The cost of labor was low, and the price for technical support was high. The supplier continued to hire numerous agents until the majority of the calls were routed to its facility. This introduced much risk into the supply chain, and the level of service decreased as the pool of educated and trained personnel started to diminish in the local marketplace.

Many factors influence the insource or outsource decision. Two primary issues are the total cost of ownership and the availability capacity. Table 3-3 lists considerations specific to each decision.

Table 3-3 Factors Influence the Insource/Outsource Decision[10]

Factors That Favor Insource (Make)	Factors That Favor Outsource (Buy)
Total cost considerations	Total cost considerations
Productive use of excess capacity	Limited production facilities
Control of production and quality	Small volume requirements
Design secrecy/patent issues	Specialized supplier expertise
Unreliable suppliers	Stability of workforce
Stability of workforce	Multiple-source policy
	Control of indirects
	Inventory considerations

The outsourcing decision is challenging with many "moving parts." There are also many cost considerations that are known and, in some respects, also hidden. In addition, the costs constantly fluctuate as economic conditions change. Fuel prices, for example, have continued to increase and caused ocean transport, inland transport, and other deliveries to increase significantly over time. The risk of terrorists after 9/11 increased security at the ports, causing the port expenses to increase.

A discussion of total cost of ownership, key in both the make-versus-buy decision and the where-do-we-buy decision, follows in the paragraphs describing some of the tools used to strategically manage costs. More discussion on total cost and geographic location decisions is also provided in Chapter 5.

Strategic Cost Management

Supply managers must understand the principles of price and cost analysis. Price is often the largest component of total cost. Hence, one of the first and most important jobs of the supply manager is to obtain the right price. However, this is even sometimes difficult to determine: What is the right price? The right price is not the same for all suppliers because of location or composition of the workforce. However, purchasers have the ability to assess the marketplace and market conditions to start to determine the right price and how it should be managed?

As supply managers are preparing to contract with suppliers, or to negotiate pricing, they have to understand how to assess the price, determine the supplier's costs, and determine the other associated costs of doing business with the supplier. This is how to strategically managed costs. The broad category of strategic cost management focuses on three particular themes and helps purchasers to appropriately manage costs across the supply chain.[11]

- **Value chain**—Managing costs requires a broad focus external to the firm (from raw material to the end use of the customer). It is different from value added analysis in the sense that it starts much early and continues throughout the entire life cycle of the product. Consider, for example, a manufacturer who decided to implement just-in-time manufacturing. This required the suppliers to ensure multiple deliveries and ultimately be responsible for the inventory. The buyer's costs were decreased because there was no more inventory in the facility. However, the supplier's transportation and inventory costs increased. As soon as it was time to renegotiate the contract, prices went up significantly with just cause. There is opportunity for savings and innovation; however, a broader perspective is necessary to think about what is good for the whole versus what is best for one.
- **Strategic positioning**—Differences exist in cost management and control depending on the strategy followed. Think about a company that practices leadership in product differentiation (BMW) versus a leader in cost management (Wal-Mart). Wal-Mart looks at ways to reduce prices (to achieve everyday low price guarantees) whereas BMW tries to understand the cost drivers and look at improved processes to make any sort of cost improvement.
- **Cost drivers (what causes cost)**—Many areas generate cost, which can be classified under structural and executional cost drivers. Structural cost drivers include things such are scale, scope, experience, technology, and complexity, whereas executional drivers include work force involvement, plant layout efficiency, product configuration, and others. Quality underlies every cost driver including prevention, appraisal, and internal and external failure costs.[12]

The broad category of strategic cost management contains these underlying themes, and some tools are used in assessment. Understanding how to apply these tools appropriately in a particular situation can help buyers to better understand the right price and help them to make better decisions related to "make" or "buy" and location.

Strategic cost management enables a purchaser to creatively manage supply chain costs while taking into account the type of commodity and the type of relationship with the supplier. There is a direct relationship between portfolio analysis and the appropriate strategic cost management techniques (Figure 3-2). Some of the tools and techniques used in strategic cost management take significant time and effort to compile and analyze the data.

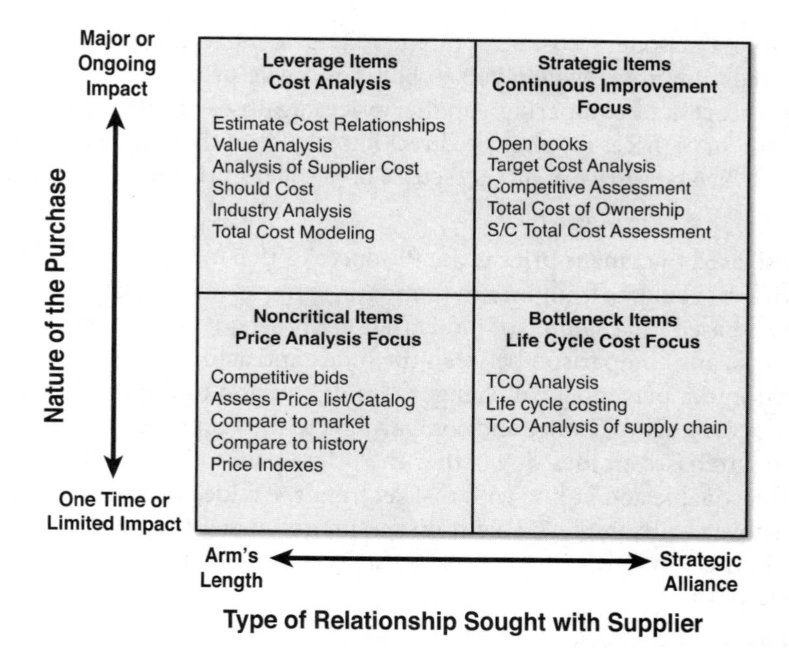

Figure 3.2 **Relationship between Kraljic's portfolio matrix and strategic cost management**[13]

Price Analysis Focus—Noncritical Items

Supply professionals understand that the price paid to a supplier is equal to cost with a profit added in (Suppliers Price = Cost + Profit). As mentioned previously, a supplier has to make a profit to stay in business. Buyers must look at a price and see if it is within the range of acceptable amounts based on current market conditions. Suppliers can underprice a bid to get the business. However, within a short time, they will ask for price increases. The cost part of this equation in broad terms consists of labor, material, overhead, and SGA (see Figure 3-3).

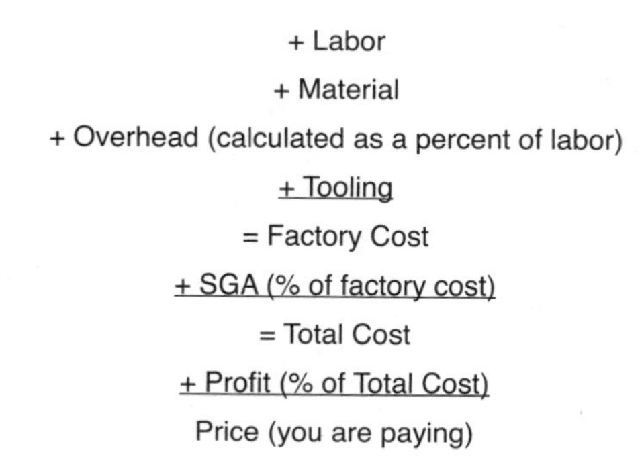

Figure 3-3 **A basic supplier cost structure**

Understanding a supplier's cost structure can help identify areas in which a supplier could potentially improve or help you to better mitigate price increases. Price analysis refers to the process of comparing supplier prices against external price benchmarks. Initial assessment of prices requires no direct knowledge of the supplier's individual cost elements such as labor. Price analysis focuses on a seller's price with little or no consideration given to the cost.

One database used to manage price is the Producer's Price Index (PPI).[14] The PPI tracks material price movements from month to month. It tracks the increase in material commodity prices based on a sample of industrial purchasers. It is an index, or a bundle of prices. To make any comparison between the index and actual prices paid, both have to be converted into a percentage of change from one period to the next. Converting the index into a percentage of change and comparing it to the supplier's percentage of change can give the purchaser an idea of whether the prices paid to the supplier over a period of time are reasonable and in line with market trends. An idea of how to perform a price analysis is shown in Example 3-1 with discussion following that you can use to assess the results. In this example, you (as a buyer) are responsible for buying fuels and related products and power.

Example 3-1: Price Analysis

Step One: Your finance/accounting department provides you the most recent actual prices paid to the supplier (see Table 3-4). It provides you with the following information from April 2003 until April of 2013. Keep in mind that you are working with indexes, and it is the percentage of change that matters, not the actual prices.

Table 3-4 Actual Price Paid: Fuel and Related Products (April 2002–April 2013)

Year	Price per Kilowatt
2003	0.07
2004	0.09
2005	0.09
2006	0.11
2007	0.12
2008	0.15
2009	0.11
2010	0.11
2011	0.15
2012	0.15
2013	0.12

Step Two: The next step in the price analysis problem is to look at the producer's price index for the commodity you are interested in.

Year	Price Per Kilowatt
2003	0.07
2004	0.09
2005	0.09
2006	0.11
2007	0.12
2008	0.15
2009	0.11
2010	0.11
2011	0.15
2012	0.15
2013	0.12

Figure 3-4 Producer's Price Index for fuel and related products and power

Step Three: Create a table (see Table 3-5) to compare the percentage of change for both of these. The percentage of change is calculated from where you started (beginning); then subtract the period of time you are considering (ending) and divide the result by where you started. This gives you a percentage of increase or decrease over a period of time. You can compare any time period of interest. For ease of explanation, only one year is compared in the table.

Table 3-5 Percent of Change Comparison Actual Versus Market

PPI				Actual	
Apr 12	Apr 13	% Change	Apr 12	Apr 13	% Change
3394	319	6.01%	0.15	0.12	20.00%

Step Four: Assess the results. Here look to see if the supplier's prices are changing in the same direction as the PPI and in the same general percentage. In the previous case both the PPI and Actual show a decrease. The actual percentage of change is greater than the market percentage of change meaning that the purchaser is doing a good job watching market trends; it potentially did some forward buying or negotiated a better price.

Look to see if the market outperforms the supplier or vice versa. You can use this assessment to determine whether purchasing teams perform well, whether an individual buyer performs well, or whether a commodity performs well. It is not uncommon for purchase price variance to be on both the supplier's scorecard and as a personal performance metric for performance appraisals.

Cost Analysis Focus—Leverage Items

More and more organizations are shifting their attention from price management to cost management. However, purchasers have to understand market trends and conditions on each commodity that they are responsible for. With cost analysis more opportunity exists to reduce costs that are not available when the discussion is based only on price. You can use cost analysis to compare and contrast quotes from different suppliers. Organizations also use this as a tool to validate supplier's claims that they need a price increase (based on the price paid) because there has been a shift in one of the cost elements (that is, labor increases).

Frequently organizations focus their supply management efforts on "clean-sheeting," which means that the buying firm develops the expected price by assessing industry standards for profit, SGA, overhead, material, and labor. They also look at pricing on the components and raw materials that go in to the product. The result is then compared to the price/cost provided by the supplier and used to negotiate price decreases.

Another situation in which cost analysis is used is when a supplier is not forthcoming about cost data. In this case, the purchaser can use "target cost analysis, should-cost, or reverse price analysis" to get an idea of whether the supplier's prices and costs are inline with expectations, and they make too much of a profit, or potentially not enough. Many companies have found great success using this technique.

A few things need additional discussion related to the cost structure (refer to Figure 3-5). First, manufacturing overhead is usually (although not always) a percentage of labor. This means that if labor changes, overhead changes. SGA represents sales and general and administrative costs and is a percentage of the total factor cost. Profit is allocated based on the total cost of the component. To verify validity of the information, many databases and industry standards exists for each cost element.

Figure 3-6 shows a side-by-side comparison of an internal cost with the quotes from two suppliers. Each of the three has different overhead ratios (as a percentage of labor), different material costs, and so on. In addition, Supplier 1 does not have the expertise to make the entire component in-house so has to subcontract the component to one of its suppliers.

Producer Price Index-Commodities
Original Data Value

Series Id: WPU057303
Not Seasonally Adjusted
Group: Fuels and related products and power
Item: No. 2 diesel fuel
Base Data: 198200
Years: 2003 to 2013

Year	Jan	Feb	Mar	Apr	May	Jun	Jul	Aug	Sep	Oct	Nov	Dec	Annual
2003	97.6	123.8	129.4	102.3	87.9	89.8	92.7	96.6	91.1	101.1	95.9	98.1	100.5
2004	109.3	103.7	109.7	119.9	121.0	114.2	123.0	135.1	140.9	166.6	159.7	135.3	128.2
2005	141.1	149.5	173.3	175.4	170.8	187.2	189.8	200.6	212.6	264.1	206.2	198.5	189.1
2006	197.1	196.2	206.5	230.4	239.6	246.9	237.5	250.2	201.3	197.5	197.2	203.0	216.9
2007	180.9	193.5	220.2	238.0	226.5	227.6	243.5	231.2	246.2	249.6	296.7	271.9	235.5
2008	278.2	287.5	353.7	385.1	398.2	421.0	431.9	346.7	342.3	281.8	224.1	168.0	324.9
2009	161.6	147.2	139.2	167.4	166.4	191.1	172.8	204.1	193.2	202.8	215.7	205.1	180.6
2010	229.4	206.9	225.5	240.0	235.8	221.8	218.5	231.1	227.7	243.7	255.3	259.2	232.9
2011	270.0	289.3	321.8	339.8	328.4	333.7	327.8	307.3	317.8	310.6	337.1	311.0	316.2
2012	322.0	329.2	344.3	339.4	325.8	295.4	298.7	324.1	342.4	351.0	323.8	317.4	326.1
2013	318.9	342.4	321.0	319.0	308.0	306.0	311.8						

Figure 3-5 A basic supplier cost structure

PPI			Actual		
Apr-12	Apr-13	% Change	Apr-12	Apr-13	% Change
339.4	319	6.01%	0.15	0.12	20.00%

Figure 3-8 Make versus buy cost comparison

Buyers would use these cost comparisons to understand "why" the costs are different, and whether those differences are materially different to the purchase. Other issues that frequently come up for purchasers are calls from suppliers that want (need) a price increase. Purchasers have to be astute enough to understand that there are different influences on each of the cost elements and then decide on the best strategy to deal with the situation.

For example, Supplier 1 from the previous example calls approximately 1 year after it starts doing business with your organization. It tells you that it is about to lose a serious labor negotiation and must raise its price by 25 percent to compensate for the expected 25 percent increase in wages. Figure 3-7 gives an example of how the cost elements are impacted by the labor problem. In this case, the supplier asked for an increase up to $10.54 from $8.43. However, assuming a 25 percent increase in labor costs, the actual increase would be to a price of $9.82 because of the way that the supplier's cost structure is built.

+ Labor

+ Material

+ Overhead (calculated as a percent of labor)

+ Tooling

= Factory Cost

+ SGA (% of factory cost)

= Total Cost

+ Profit (% of Total Cost)

Price (you are paying)

Figure 3-7 Using a cost analysis to look at potential price increase

Total Cost of Ownership—Strategic and Bottleneck

Total cost of ownership (TOC) is a technique to help manage costs across the supply chain. Total cost analysis goes beyond purchase price, transportation, and tooling to encompass many other aspects of the cost of doing business with a supplier. TCO also allows a purchaser to run a "what-if" (or a sensitivity) analysis on key cost elements. This can be especially useful if there is a potential offshore purchase. The cost elements are often broken down into four broad categories (Table 3-6).

Table 3-6 Total Cost of Ownership: Cost Elements

Cost Element	Description
Purchase price	Often (but not always) the biggest part of the total cost.
Acquisition costs	All costs associated with bringing the product, service, or capital equipment to the customer's location. Examples of acquisition costs are sourcing, administration, freight, and taxes.
Usage costs	This involves the costs of converting the component into a finished good and supporting it through its usable costs: inventory, conversion, scrap, warranty, training, downtown, and so on.
End-of-life	All costs incurred when a product, service, or capital equipment reaches the end of its usable life: disposal, reverse logistics, clean-up, salvage, and so on.

The most challenging part of a total cost of ownership analysis is determining what cost elements are materially relevant to this product and then gathering data for each cost element. This process involves many different functions of the organization. Developing a total cost model involves an understanding of the entire process for a particular item.

When putting together a total cost model, one important thing is to first start with the price of the item and then list all the potential cost elements that may be involved throughout the life of the product. A lot of brainstorming and data collection needs to involve many members of the organization.

You want to understand how much it costs to do business with a particular supplier or in a particular region of the world. You can use total cost models to better understand the risks associated with doing business with a supplier. Table 3-7 is an example of a total cost model used to compare and contrast four different areas of the world in the purchase of small engines.

You can look at many things when using this analysis. First, the lowest price/total for one of these items is not necessarily the best for your organization. In this case, it is China that has the lowest price. The buying company has just started to make a push toward implementing a sustainable sourcing strategy and finds out that China's practices are not as good as this company needs. The United States is the most expensive: Does it make sense to buy in the United States? It appears that there is much less risk associated with that purchase.

Table 3-7 A Total Cost of Ownership Model for a Small Engine

	China (Container)	Eastern Europe (Container)	Western Europe (Container)	U.S. (each)
Unit Price	$42,000.000	$60,000.000	$85,000.000	$120.000
Packaging	$2,000.000	$2,500.000	$1,750.000	$1.000
Tooling	$100.000	$75.000	$125.000	$0.025
Inland Transportation	$200.000	$750.000	$200.000	$5.200
Ocean Transportation	$3,000.000	$4,500.000	$3,500.000	
Freight Fwder Fee	$200.000	$400.000	$100.000	
Insurance	$420.000	$600.000	$850.000	
Brokerage Fee	$400.000	$400.000	$400.000	
Travel	$250.000	$250.000	$250.000	
Administrative	$150.000	$200,000	$75,000	
U.S. Port Handling Charge	$1,500.000	$1,500.000	$1,500.000	
Duty	$2,100.000	$3,000.000	$4,250.000	
Brokerage 2	$500.000	$500.000	$500.000	

	China (Container)	Eastern Europe (Container)	Western Europe (Container)	U.S. (each)
Port to Avondale	$1,500.000	$1,500.000	$1,500.000	
Cost of Capital	$840.000	$1,200.000	$1,700.000	
Warehouse Costs	$1,600.000	$1,600.000	$1,600.000	
Defect Rate	$2,100.000	$1,800.000	$850.000	
Environmental Surcharge			$500.000	
Environmental Surcharge			$750.000	
Price for Container	$58,860.000	$80,775.000	$105,400.000	
Price Each	$58.860	$80.775	$105.400	$126.230

The issue is that if you don't go with the lowest price, as a purchaser you have to justify your position. You must think about how to build your business case. Think about the strategic concerns of an organization, for example, environmental issues. Consider if there is volatility in some of the cost elements: maybe freight or port handling? For most of the items, you can run a what-if (sensitivity) analysis to see what happens at a certain level of increase.

The other main function of these total cost models is to allow the purchaser to look at the high cost elements and creatively think of a way to reduce those. The whole process involves information gathering, brainstorming, and a key commitment to assess the supplier's use of this type of model. Total costs are discussed more in Chapter 4 and Chapter 5.

Purchasing and Supply Chain Analytics

Buyers can use a number of other analytical tools to assess their suppliers and supply chain operations. These are in no way all-encompassing but provide additional "tools" to put in your toolkit to help in decision making.

Break-Even Analysis

Break-even analysis includes both cost and revenue data for an item and identifies the point in which revenue equals cost, and the expected profit or loss at different production levels.[15] Break-even analysis is used for supply management to identify if a target purchase price gives a supplier a reasonable profit. Suppliers often underbid a price to qualify for the contract. However, long-term viability of suppliers depends on profit.

Break-even analysis also helps to analyze a supplier's cost structure. At the least understanding the fixed and variable costs involved in a supplier's operation is important. However, as previously noted understanding the cost elements is key to making an effective decision. Buyers can perform a what-if analysis using a break-even analysis. They may look at different mixes of product volumes and purchase price.

Break-even analysis also helps a buyer to prepare for negotiation and anticipate a seller's pricing strategy. A direct relationship exists between preparation and success of the negotiation.

The break-even analysis begins with identifying the price (P), expected volume (X), fixed costs (FC), and variable costs (VC). Then use Net Income or loss = P (X) – VC (X) – FC. You can also determine how many units the supplier needs to produce to break even using P (X) = VC (X) + FC. Manipulating the different variables can help you to understand what might happen "if."

Learning Curve Analysis

Learning curve analysis is the rate of improvement due to learning as producers realize a direct-labor cost improvement in production. The rate of learning can help during a negotiation especially as it relates to the length of the contract. Also, learning curve rates are calculated using direct labor, which is relatively easy to access. Table 3-8 shows the rate of learning and how it is calculated.

Table 3-8 Learning Curve with a Rate of Learning of 80 Percent

Labor	$12.0	$9.6	$7.7	$6.1	$4.9
Material	$8.0	$8.0	$8.0	$8.0	$8.0
Overhead (110%)	$13.2	$10.6	$8.4	$6.8	$5.4
Total Factory Cost	$33.2	$28.2	$24.1	$20.9	$18.3
SGA (10% of Factory Cost)	$3.3	$2.8	$2.4	$2.1	$1.8
Total Cost (per Unit)	$36.5	$31.0	$26.5	$23.0	$20.2
Total Units	$2,500.0	$5,000.0	$10,000.0	$20,000.0	$40,000.0
Total Contract Cost	$91,300.0	$154,880.0	$265,408.0	$459,852.8	$806,164.5

Notice in the previous example that each time production doubles, goes from 2,500 to 5,000 units, for example, labor decreases by 20 percent (1–80 percent learning curve). The total contract cost is the number of units times the cost per unit. Realize too that in reality the more a supplier performs this particular task, the more likely other costs will drop as well. The learning curve effect is not just restricted to labor.

You can use the results of this assessment in negotiations, understanding that the more products a supplier produces, the better it should be at producing; and therefore the price should go down. The learning curve is a powerful but relatively simple tool to use.

Quantity Discount Analysis

This tool is used to understand the incremental changes in cost between quantities within a supplier's price quotation. QDA allows the user to verify that quantity discounts are reasonable. Using this allows a buyer to negotiate price improvements (Figure 3-8).

	Supplier 1	Supplier 2	In-House
+ Labor	$ 2.00	$ 2.50	$ 2.25
+ Material	$ 1.00	$ 1.50	$ 1.25
+ Subcontracted Components	$ .50	$	
+ Overhead (calculated as a percent of labor)	$ 2.80	$ 3.13	$ 3.38
+ Tooling	$ 1.00	$ 1.50	$ 0
Factory Cost	$ 7.30	$ 8.63	$ 6.88
+ SGA (% of factory cost)	$ 0.73	$ 0.43	$ 1.03
Total Cost	$ 8.03	$ 9.06	$ 7.91
+ Profit (% of Total Cost)	$ 0.40	$ 0.23	
Price (you are paying)	$ 8.43	$ 9.28	

Figure 3-8 Quantity discount analysis

Analysis of the price breaks often reveals an up-and-down roller-coaster effect between incremental price differences. If buyers can learn to ask questions about the discounts, they may achieve even greater savings.

Service Purchase Analysis

Purchasers have an opportunity to become involved in better managing the services spend. Many organizations have started to audit (or hire consultants to audit) categories such as temporary labor spend, legal spend, travel, and many others. As purchasers engage in this type of analysis, they realize that they can achieve significant savings. Table 3-9 shows a fictitious audit of temporary labor spend. The end result of this analysis is that overbilling by the temporary labor firm represented approximately 1 percent of total revenue that was lost.

Table 3-9 Implications of Poor Services Purchasing[16]

Implications of Improper Management of Services (000's)	
$10,000,000	Total Revenue
$1,460,000	Services Spending (14.6% of Total Revenue)
$90,520	"Unearned" Excess Services Billing (6.2% of Services Spending), equivalent to lost profit
$2,000,000	20% Profit Margin 4.53% Profit Reduction due to Excess Services Billing (90,520/2,000,000)

A few of the many types of supply chain analytics can be performed to ensure that as a buyer you are doing the best you can to protect you company from financial risk. Buyers need to be well versed in both the analytical skills and the relational skills needed to work with suppliers.

Sustainability and Ethics

The final section of this chapter introduces the idea of sustainable supply chain management. In addition, purchasers must understand that because of the significant amount of money that purchasing is responsible for, an ethical code of conduct was developed to ensure that the money is spent in a way that keeps the buyer and the buying organization out of some difficult positions preventing tarnished reputations.

Sustainability

Environmental sustainability can be defined as

> Using the earth's resources in such a way to meet the needs of the present without compromising the ability of future generations to meet their own resource needs.

Most environmental initiatives that involve purchasing and its suppliers are in the early stages. Although environmental metrics are being incorporated into requests for proposals, they are often the order qualifier, not the final deciding factor for which supplier will receive the bid. What appears to be an increasing trend is that buyers are asking for specific environmental and social initiatives to come from the supplier. More often buyers will ask for supplier's to be environmentally certified. In some cases, the supplier uses environmental initiatives and objectives as a way to gain business from the buyer. Innovative environmental ideas in the areas of waste reduction, GHG reduction, and energy reduction are becoming part of everyday business conversation between buyer and supplier.

Packaging suppliers are often tasked to work with packaging engineers to reduce the amount of materials headed to the landfill. FedEx has been working with its suppliers on changing packaging from "virgin" to recycled material.[17] Baxter worked with FedEx to develop an alternative to shipping its delicate products by air, saving on both emissions and cost.[18]

In addition, many opportunities exist for purchasers to set expectations and introduce environmentally sustainable metrics for their suppliers—purchasers can learn from the supply base. In many cases, their suppliers are already introducing environmental metrics into their processes to meet the needs of their other customers. Purchasing has a key role to help an organization meet its sustainability goals and implement these goals beyond the walls of its own facilities out to the supply chain.

Purchasing has four key roles that it can perform to start to incorporate the suppliers in its initiatives.

- Gather data and manage suppliers to support regulatory compliance efforts.
- Communicate expectations to first-tier suppliers (and critical second-tier suppliers, if necessary).
- Measure first-tier suppliers' performance in this area.
- Identify and implement specific projects.

When purchasing optimizes and rationalizes its supply base, it may also realize that there are many suppliers remaining that already have excellent environmental programs in place. These suppliers also have a number of excellent ideas for projects and initiatives that they can help to implement. Chapter 6 talks about the different trade-offs that purchasers have to make in their buying decisions. These trade-offs used to be cost, quality, and service. Now, carbon is added in to the mix, making it even more challenging to make certain buying decisions.

Supply Management Ethical Code of Conduct

The Institute for Supply Management developed a code of conduct for those in the supply management organization. This code of conduct has been adopted and adapted by many organizations to ensure that the buyers act in a responsible way. The code of conduct has three overarching principles that supply managers must adopt.

- Integrity in your decisions and actions
- Value for your employer
- Loyalty to your profession

From these principles 10 standards of supply management conduct were developed (see Table 3-10). For more specific information on ethical conduct please see http://www.ism.ws/files/SR/PrinciplesandStandardsGuidelines.pdf.

Table 3-10 ISM's Ethical Standards[19]

Impropriety	Prevent the intent and appearance of unethical or compromising conduct in relationships, actions, and communications.
Conflict of interest	Ensure that any personal, business, and other activities do not conflict with the lawful interests of your employer.
Issues of influence	Avoid behaviors or actions that may negatively influence, or appear to influence, supply management decisions.
Responsibilities to your employer	Uphold fiduciary and other responsibilities using reasonable care and granted authority to deliver value to your employer.
Supplier and customer relationships	Promote positive supplier and customer relationships.
Sustainability and social responsibility	Champion social responsibility and sustainability practices in supply management.
Confidential and proprietary information reciprocity	Protect confidential and proprietary information. Avoid improper reciprocal agreements.
Applicable laws, regulations and trade agreements	Know and obey the letter and spirit of laws, regulations, and trade agreements applicable to supply management.
Professional competence	Develop skills, expand knowledge, and conduct business that demonstrates competence and promotes the SM profession.

Conclusion and Chapter Wrap-Up

This chapter contains a lot of information primarily because of the changing role of supply management into a more strategic contributor. Category sourcing strategies were introduced again so that purchasers can understand how to use the portfolio matrix to implement strategy and also to link to strategic cost management. Other strategies include supply base rationalization and optimization. Purchasers have to know the right number of suppliers to have a highly effective supply base. There was a brief discussion on insource versus outsource.

The chapter progressed into a more analytical phase of the sourcing process. Tools can help supply managers manage suppliers' bids and proposals. These tools can also help buyers to be better prepared as they enter into the negotiation process.

Finally, the importance of sustainability and ethics in supply management is introduced. Both of these ideas are key themes that incorporate all the aspects of the purchasing process.

The key points from this chapter include the following:

- Develop a more in-depth understanding of the key supply management principles and strategies.
- Work more with portfolio analysis to understand how the sourcing strategy for each of the procurement categories relates to organizational strategies.
- Learn how and why a supply base is rationalized and optimized.
- Determine the role of supply management in the insource (make) versus outsource (buy) decision and the associated risks and benefits of each.
- Develop the capability to understand and prepare the appropriate cost management analyses and other supply chain analytics.
- Discover some of the keys to supply management involvement in environmental sustainability and also in ethical purchasing practices.
- The final key takeaway from the chapter is that it takes both highly fine-tuned analytical skills and excellent communication and relational skills to be successful in supply management.

Key Terms

- Supply management
- Sourcing strategy
- Strategic supplier
- Transactional supplier
- Supply base rationalization and optimization
- Strategic cost management

- Price analysis focus
- Cost analysis
- Total cost of ownership
- Break-even analysis
- Learning curve analysis
- Quantity discount analysis
- Service purchase analysis

References

1. Monczka, et al., (2011). *Purchasing & Supply Chain Management.* Southwest Publishers: Mason, OH.
2. Moore, et al., (2002). "Implementing Best Purchasing and Supply Management Practices." Lessons from Innovative Commercial Firms. Documented briefing sponsored by Rand and compiled for the U.S. Air Force.
3. Monczka, et al. (2011), ibid.
4. Adapted from Monczka, et al. (2011), ibid.
5. ISM (2005). "Spend Analysis and Supply Base Rationalization", Institute for Supply Management. Retrieved on September 1, 2013, from http://www.ism.ws/files/tools/spendanalysissupplybase.pdf.
6. Patterson, S. (2005). "Supply Base Optimization and Integrated Supply Chain Management." Contract Management. Retrieved on September 1, 2013, from http://www.ncmahq.org/files/Articles/F4CB4_CM_Jan05_p24.pdf.
7. Adapted from Burt, Dobler, Starling (2005); Monczka, et al. (2011); Benton (2007); and Johnson, Leenders, Flynn (2006).
8. Martin, A. (2013). "Why Rationalize Your Supplier Base?" Ezine Articles. Retrieved on September 1, 2013, from http://ezinearticles.com/?Why-Rationalize-Your-Supplier-Base&id=332671.
9. Adapted from Patterson, S. (2005), ibid.
10. Monczka, et al (2011), ibid.
11. Shank, J. and Garvindarajan, 1993.
12. Uran, the costs of quality.

13. Ellram article on strategic cost management.
14. Bureau of Economic Analysis, Producers Price Index, can be accessed from www.bls.gov.
15. Investopedia (2013). What Is Break Even Analysis? Retrieved on September 17, 2013 from http://www.investopedia.com/terms/b/breakevenanalysis.asp.
16. Ellram, Tate, Billington (2009). Services: "The Last Frontier." *California Management Review*.
17. GEMI (2004). Creating Environmental Value at FedEx Express. Retrieved on September 17, 2013, from http://www.gemi.org/supplychain/g1e.htm.
18. Baxter (2012). "Decreasing Environmental Impact of Product Transportation." Retrieved on September 15, 2013 from http://sustainability.baxter.com/quick-links/case-studies/2012-report/product-transport.html.
19. ISM, "Ethics and Business Conduct." Retrieved on September 13, 2013, from http://www.ism.ws/SR/content.cfm?ItemNumber=4762.

4

THE CRITICAL ROLE OF TECHNOLOGY IN MANAGING SUPPLY OPERATIONS AND PRODUCT FLOWS

This chapter describes the role of technology in supporting supply management operations. The goal of this chapter is to identify the key categories of supply management software and their capabilities, and discuss the emerging technology solutions and capabilities in supply operations.

Learning Objectives

After completing this chapter, you should be able to:

- Discuss the role and capabilities of technology in sourcing.
- Identify some of the different types of technology.
- Examine some of the trends that are impacting supply management operations.

Technology in Supply Operations

Supply managers today expect and rely on powerful technological solutions to their business problems. However, these sophisticated systems and solutions have not always been available and continuously evolve. Most of the early uses of technology were in the accounting and financial areas, primarily focusing on keeping track of revenue and ensuring that bills to suppliers were paid on time. Beginning in the early 1970s, information technology resources and software systems were developed and designed to help facilitate supply chain operations.

Materials Requirements Planning (MRP) and Distribution Requirements Planning (DRP)

Early supply chain technological solutions were primarily focused internally with a key purpose of helping to manage the ever-increasing level of inventory. Knowing what was available and where it was located in the facilities helped logistics and manufacturing managers better meet shipping and production schedules. These systems included technology solutions such as material requirements planning (MRP) and distribution requirements planning (DRP).[1]

MRP was a computerized ordering and scheduling system primarily for the manufacturing industries. It uses bill-of-material data, inventory data, and the master production schedule to project what material is required, when, and in what quantity.[2] MRP is a plan-push system and is a backward scheduling system. This means that plans are generated and then the requirements are cascaded down through the bill-of-materials all the way to the purchase requisition. It takes into account existing inventory both in raw material and work-in-process.

MRP transitioned to a more advanced Manufacturing Resource Planning system (MRP-II). This technology incorporated planning of all aspects of a manufacturing firm. This includes business planning, production planning and scheduling, capacity requirement planning, job costing, financial management, and forecasting. [3] Decisions made in MRP-II were more informed decisions than in MRP-I, but these decisions were still inhibited and not communicated out of the supply chain.

DRP and MRP were linked together. DRP was the process for determining which goods, in what quantities, at which location, and when they are required to meet anticipated customer demand. The inventory related information from the DRP is input into the MRP system as gross requirements to estimate input flows and production schedules.[4]

With these two systems, there was still no connection to the supply or customer base making it difficult to effectively plan requirements outside of the manufacturing area. Conflicts between the marketing and sales area, manufacturing, and purchasing were common because of the inability to accurately share information. There was little collaboration and planning was independently performed to optimize a single function. Customer requirements were met but often at the expense of additional inventory and higher costs.

Electronic Data Interchange (EDI)

The early technology solutions were followed by more advanced systems that enabled the electronic transfer of customer and supplier information called electronic data interchange (EDI). Today, EDI is primarily Internet-based (web-based) but still used by many organizations.

Initial EDI implementations were largely driven by the retail customer to help better facilitate the flow of orders. EDI was the initial foray into the highly advanced point-of-sale systems that frequent retail marketplaces today.[5]

Initial EDI systems were costly to implement, and it took quite a while for manufacturers to adapt to the systems. Although, now because of improvements and the move to Internet-based systems, even smaller businesses can afford the EDI infrastructure.

There were many hurdles to implementation in the early stages of EDI. Actually, there were often reports of data being transferred electronically from the retailer to the manufacturer, but at the manufacturer the systems couldn't accept the data electronically and transfer it to the required databases. The manufacturers would consolidate the data that was electronically transmitted and manually enter it into the system, losing the advantage of the electronic transmission.

Now, the benefits of EDI are numerous. EDI was the stepping stone to some of the more advanced technologies used today. EDI saves money and time because transactions can be transmitted from one information system to another through a telecommunications network, eliminating the printing and handling of paper at one end and the inputting of data at the other, which as mentioned was common in the initial stages of implementation.[6]

EDI was developed to solve the problems inherent in paper-based transaction processing and in other forms of electronic communication. EDI helped resolve some of the issues associated with paper-based transactions including time delays, high administrative costs, accuracy of data, and access to information.

Enterprise Resource Planning (ERP)

Enterprise resource planning (ERP) systems then entered the marketplace. Companies such as Oracle, SAP, JD Edwards, and PeopleSoft dominated the market in ERP sales and implementation.[7] The goal of ERP is to integrate all business function planning and processing.[8] ERP systems take more of a siloed business approach for each of the major business functions and deposit data in a centralize database (Figure 4-1). For example, there are *modules* for materials management and production planning. ERP systems made it easier to track the work-flow across various departments and reduce the operational costs involved in manually tracking and perhaps duplicating data using individual and disparate systems.[9] ERP systems have a number of advantages but also some disadvantages that may inhibit implementation (Table 4-1).

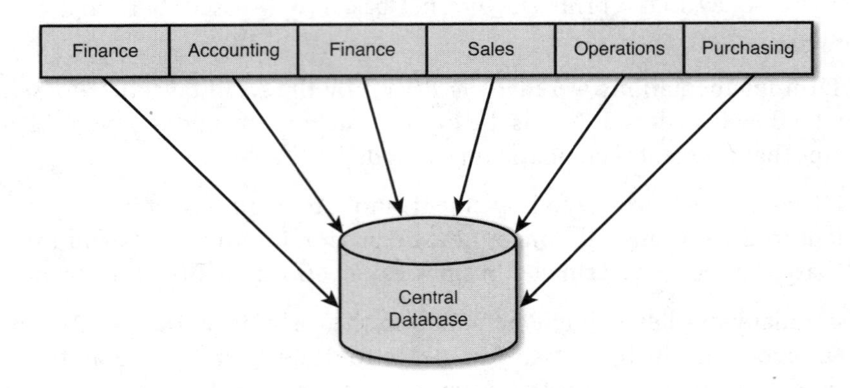

Figure 4-1 ERP system architecture

Table 4-1: Advantages and Disadvantages of ERP Systems[10]

Advantage	Disadvantage
Visibility	Cost
Collaboration across department	Time to implement
Automatic and coherent work flows.	Difficulty and cost to customize
Unified and single reporting system	Lengthy and difficult to measure ROI
Single software system for all departments	Requires extensive user training
Security	Migration of data
Ability to link other systems (that is, bar code)	
Business intelligence	
e-commerce integration	

ERP systems are primarily internally focused, as a way to run the business, and lack the supplier and customer interface. However, one of the advantages of an ERP system is that advanced e-commerce integration is possible with those systems that can handle web-based order tracking and processing. This was one of the early steps on the path to e-sourcing.

It was the influx and power of the Internet that provided the bridge between organizations and their suppliers and customers into the ERP systems. A new technology known as *e-sourcing* was developed (and is developing) with the functionality to better manage both the supply base (supplier relationship management [SRM]) and the customer (customer relationship management [CRM]).[11]

SRM systems were designed to boost contract spending and enforce compliance with procurement guidelines. [12] Supplier relationship management software helps organizations to automate, simplify, and accelerate the business' procure to pay processes for goods and services. It also helps to reduce purchasing costs by closing the loop from the sourcing period to supplier payment and certainly in the administration of the purchasing process. It automates operational processes to reduce purchasing errors, eliminate manual tasks, and helps avoid maverick buying. SRM was also another way to build collaborative supplier relationships and integrate suppliers into the procurement process.

CRM is a business strategy aimed at understanding, anticipating, and responding to a prospect or customer's needs. A primary objective of a CRM system is to provide a single, consolidated, and holistic view of the customer relationship across the organization. CRM also helps better manage the communication and interactions with customers. Finally, the customer relationship management system helps to synthesize, analyze, and understand customer data to predict customer behaviors, identify customer opportunities, or better serve customers, which helps in the upstream planning process.[13] CRM systems help to minimize the bullwhip effect.

Technology Today

Technology today is aimed at collaboration both internally and among supply chain partners, customers, and suppliers. Point-of-sale systems, RFID, product life cycle software, bid optimization, and computerized negotiation tools are now available to help facilitate the buy-side transactions with the sell-side transactions.

These new systems are also linked to the e-mobile environment with tablets and cell phones; supply managers access data on a 24/7 basis all around the world. Social networking, normally thought of as a way to connect individuals, is slowly finding its way into the B2B marketplace. Supply managers can expand their relationships both within organizations and in the general buying community. Sites such as LinkedIn help to facilitate the development of relationships between buying and selling organizations.

Case in Point—Wal-Mart and Technology

Wal-Mart has become the world's largest and most powerful retailer with the highest sales per square foot, inventory turnover, and operating profit of any discount retailer.[14] Wal-Mart began with the goal to provide customers with the goods they wanted when and where they wanted them. Next, it focused on developing cost structures that allowed it to offer low, everyday pricing. The replenishment of inventory is one of the key facilitators behind this strategy.

Wal-Mart's success is primarily due to efficient integration of supplier, manufacturing, warehousing, and distribution to stores. There are four key components of its supply chain strategy: supplier partnerships, cross docking and distribution management, technology, and integration.[15]

Technology plays a key role in Wal-Mart's supply chain serving as the foundation of its efficient and effective supply chain. Wal-Mart has the largest information technology infrastructure of any private company in the world. Its state-of-the-art technology and network design enable Wal-Mart to accurately forecast demand, track and predict inventory levels, create highly efficient transportation routes, and manage customer relationships and service response logistics.

Wal-Mart designed a point-of-sale system that improved customer services, decreased inventory throughout its supply chain, and improved the overall competitive capability of the company. EDI is the underlying technology that enables the system to work.

There are a number of benefits to Wal-Mart's supply chain partners. For suppliers, demand is visible, obsolete stocks are minimized, volume increases, capacity planning is improved, and so on. This networked supply chain process has helped to change the way consumer-products companies and wholesaler distributors do business with retailers.[16]

E-Sourcing

E-sourcing is the application of technology tools to the strategic sourcing process to enable faster cycle times, improve cost performance, and increase competitive advantage by reducing supply redundancies, increasing speed and flexibility, and maximizing the combined organizational benefits of centralization and decentralization.[17]

The business-to-business market is large with many opportunities to incorporate e-sourcing tools. E-sourcing enables a company to take advantage of Internet-enabled purchasing tools to improve the efficiency and effectiveness of its overall spend. E-sourcing takes any number of forms from buy-side and sell-side e-catalogs to electronic request for quote (RFQ) procedures that allow purchasers to post specifications and solicit bids to cyberspace commodity exchanges where buyers and sellers meet and trade.[18]

E-sourcing does more than establish an electronic venue for buyers and sellers to meet. It also streamlines workflows, enhances flexibility, and drives transparency in the buyer-seller relationship. E-sourcing improves the accuracy and availability of information on both the supply and demand side, facilitating collaboration, control, and compliance. That type of knowledge makes for more informed negotiations and richer arbitrage opportunities.[19] Also, by eliminating routinized tasks like transaction processing, e-sourcing can free up purchasing personnel to focus on more strategic issues like comprehensive supplier screening, supply base development and relationship management, linking suppliers into innovation processes, and others. There are many benefits of e-sourcing both operational and strategic (Figure 4-2).

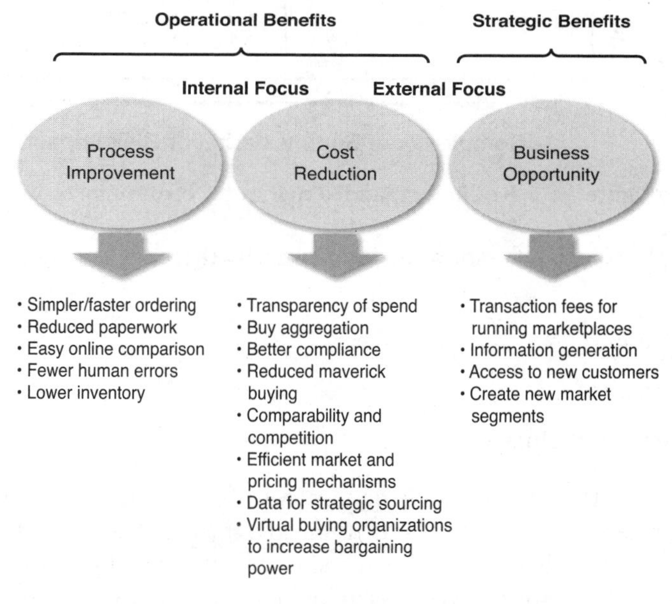

Figure 4.2 Benefits of E-sourcing tools[20]

The applicability of e-sourcing tools varies according to the type of product or service. Going back to Kraljic's (1983) matrix and applying e-sourcing tools and solutions to particular types of commodities indicates that both the objectives and the solutions vary (Figure 4-3).

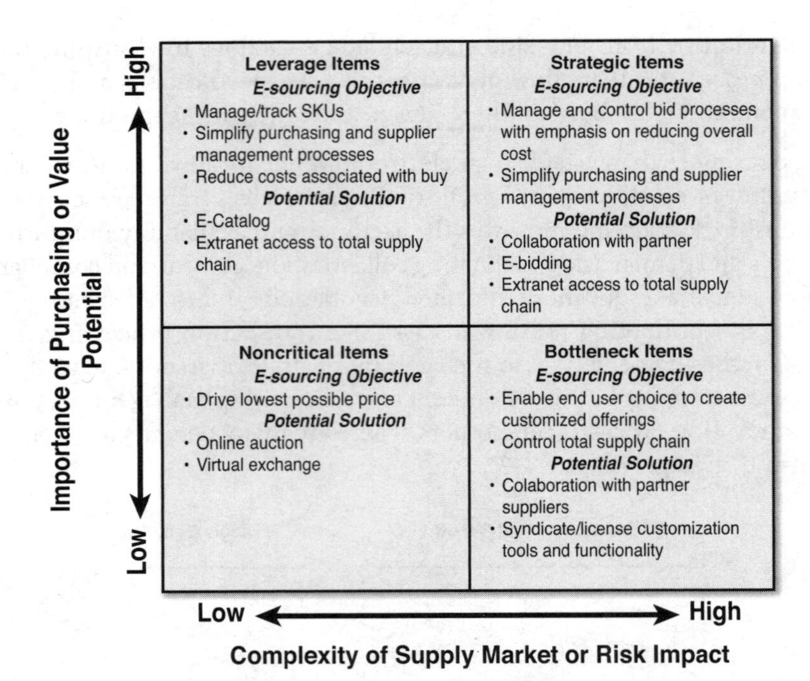

Figure 4-3 Kraljic's expanded matrix for e-sourcing tools

There are multiple aspects of e-sourcing but primarily three types of e-sourcing business models[21]

- Sell-side venues
- Buy-side venues
- Third-party marketplaces

Sell-side systems contain the products or services of one or more suppliers. A site like alibaba.com represents a sell-side system. Typically, the registration to a sell-side system is free to users. The supplier guarantees the site security, and there is no investment by the buyer. These systems are easily accessed by many different suppliers. However, there is an inability to track expenditures or to control expenditures.

General Mills established a supplier's training manual for using the e-sourcing sell-side system.[22] Suppliers can use the system to schedule meetings, review scorecards and other performance metrics, view and respond to public auctions, and requests for information and proposals. The supplier receives an invitation via email that contains a hyperlink. Suppliers then log into the system and indicate whether they are going to respond, and find out specific requirements. During the process of responding, they can "chat" with the buyer (and others). The system enables buyer-supplier interaction and real-time updating. The systems are user-friendly with technology support available when needed.

Buy-side systems are controlled by buyers. Buy-side systems enable the supply manager to manage the entire sourcing cycle, track spend, and effectively manage contracts. These may be either self-designed or provided by e-sourcing suite suppliers. They give supply managers the ability to manage the sourcing cycle, track spend, and exert control of contract management. These systems require an initial investment.

Third-party marketplaces are firms that neither buy nor sell but seek to facilitate the electronic purchasing process for value enhancement. Some vertical portals specialize in a commodity (that is, steel or chemicals). Alternatively, horizontal portals provide a broad base of products and services. The use of these portals faded as many firms developed their own web presence through sell-side sites (for example, homedepot.com and walmart.com).

One recent trend in the retail sector is to allow the capability for other suppliers to sell their products/services on their site. For example, Wal-Mart allows the customer to buy directly from other suppliers on its website. Wal-Mart increases its online sales through the e-business portal, and the supplier gains increased exposure.

Figure 4-4 shows other aspects of e-sourcing and includes different types of catalog options and the degree of interaction with each aspect. E-sourcing tools help to streamline processes and leverage technology to meet the needs of the organization. Supplier Relationship Management (SRM) software enables the purchaser to manage the entire purchasing cycle from recognition of need through contract management and supplier evaluation.

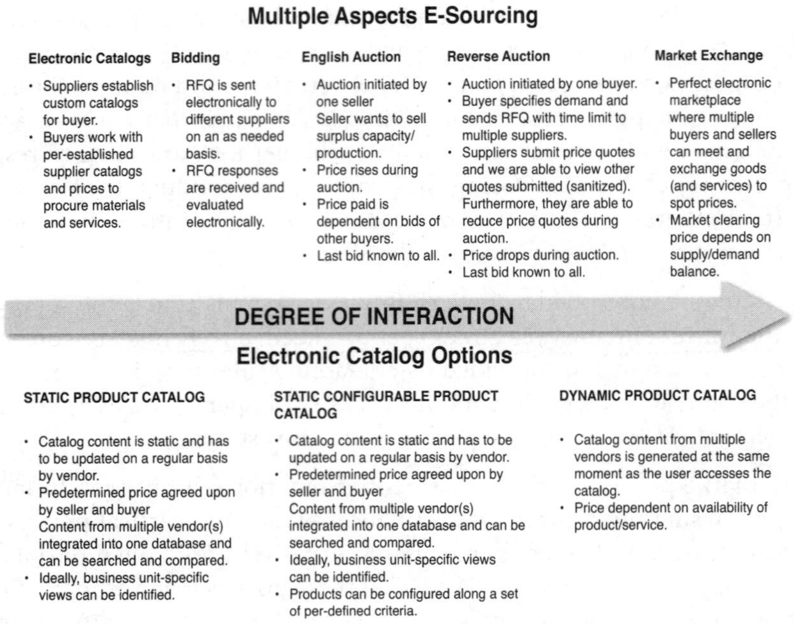

Multiple Aspects E-Sourcing

Electronic Catalogs

- Suppliers establish custom catalogs for buyer.
- Buyers work with per-established supplier catalogs and prices to procure materials and services.

Bidding

- RFQ is sent electronically to different suppliers on an as needed basis.
- RFQ responses are received and evaluated electronically.

English Auction

- Auction initiated by one seller
  Seller wants to sell surplus capacity/ production.
- Price rises during auction.
- Price paid is dependent on bids of other buyers.
- Last bid known to all.

Reverse Auction

- Auction initiated by one buyer.
- Buyer specifies demand and sends RFQ with time limit to multiple suppliers.
- Suppliers submit price quotes and we are able to view other quotes submitted (sanitized). Furthermore, they are able to reduce price quotes during auction.
- Price drops during auction.
- Last bid known to all.

Market Exchange

- Perfect electronic marketplace where multiple buyers and sellers can meet and exchange goods (and services) to spot prices.
- Market clearing price depends on supply/demand balance.

DEGREE OF INTERACTION

Electronic Catalog Options

STATIC PRODUCT CATALOG

- Catalog content is static and has to be updated on a regular basis by vendor.
- Predetermined price agreed upon by seller and buyer
  Content from multiple vendor(s) integrated into one database and can be searched and compared.
- Ideally, business unit-specific views can be identified.

STATIC CONFIGURABLE PRODUCT CATALOG

- Catalog content is static and has to be updated on a regular basis by vendor.
- Predetermined price agreed upon by seller and buyer
  Content from multiple vendor(s) integrated into one database and can be searched and compared.
- Ideally, business unit-specific views can be identified.
- Products can be configured along a set of per-defined criteria.

DYNAMIC PRODUCT CATALOG

- Catalog content from multiple vendors is generated at the same moment as the user accesses the catalog.
- Price dependent on availability of product/service.

Figure 4-4 E-sourcing and catalog options[23]

Capabilities of Supplier Relationship Management e-Sourcing Suites

An SRM e-sourcing module supports many of the sourcing processes discussed in previous chapters. The idea is that SRM systems enable purchasers to make improved decisions because of access to better data and more information. This includes supplier selection, contract management, contract compliance, and so on. The SRM modules act as an interactive system that supports the purchasing manager.

Spend analysis (discussed in Chapter 2) is a key benefit of e-sourcing. After the baseline is established through consolidation of purchase orders and accounts payable records, purchasers can look at opportunities to reduce spend through

- Consolidation of similar purchases
- Reduction in the number of suppliers
- Reduction of maverick spend
- Reduction of spend by other departments (Human Resources, marketing, and finance)

- Developing more efficient contracting methods
- Use of contracting methods to reduce risk and increase supply assurance
- Compliance to contract rate and off-contract spend

Sourcing is discussed in Chapters 1, 2, 3, and 5. Sourcing modules can help to facilitate decision making through many of the steps in the sourcing process. The idea is to reduce the time required for the sourcing process and increase the efficiency associated with processing transactions. Some of the areas that are improved in using e-sourcing tools follow:

- **Request for quotation (RFQ)**—A document that an organization submits to one or more potential suppliers to elicit quotations for a product or service. Typically, an RFQ seeks an itemized list of prices for something that is well-defined and quantifiable.
- **Bid optimization**—The use of software (and the software itself) to analyze complex buys during strategic sourcing. The result is a plan that meets the price, quality, delivery, and service needs of the buyer, while allowing the supplier to leverage its preferred offerings.
- **Reverse auctions**—A tool used to drive price down through competitive bidding. The bidders are generally prequalified in some way, and then after the opening bid, the price gets lower until the final bid wins.
- **Negotiations**—Planning for a negotiation requires data and market analysis. E-sourcing supports this.
- **Total cost support**—As mentioned in earlier chapters, total cost analysis requires a significant amount of data collection. E-sourcing tools can help to support total costs.
- **Receiving and inspection**—Tools that allow for quality and quantity verification upon the receipt of a shipment.

Contract management and compliance are discussed in Chapters 2 and 5. Contract management is one of the more challenging aspects for supply managers. Contract management processes ensure that the corporation is using standard terms, that risks are mitigated, and that the appropriate contract is in place for all the key relationships. Contract management e-sourcing tools help in contract authoring, negotiation, and approval. They also help with contract administration, enterprise reporting, and contract control. Payment obligations are managed and profit opportunities are visible. Contract management e-sourcing tools enable more transparency in the flow of information across business units and inter-organizationally.

Following are a few critical capabilities needed for an effective contract management e-sourcing tool.[24]

- **Searchable centralized repository**—This is the capability to maintain and track all the organization's contracts. These repositories should be available to any member of the organization that has the appropriate security level.
- **Collaborative capabilities and basic workflow**—This capability makes it so that multiple users are allowed to input, update, track, and create contracts and associated purchases simultaneously. The status of a contract can be tracked at any given point in time.
- **Monitors and alerts**—Users can define these monitors and alerts with upcoming contract expirations, automatic renewals, required approvals, and detected deviations. Alerts can be set up to remind a purchaser to verify budget compliance, ensure appropriate supplier monitoring, or develop schedules for supplier development.
- **Reporting and analytics**—The reports and analytics should be easy to navigate and be flexible.
- **Import/export capabilities and fixed point integration**—The manual entry of thousands of contracts is time-consuming and costly. This capability helps to eliminate that and minimize user input errors.
- **Roles, permissions, and security**—The e-sourcing contract management system should be accessible and usable by everyone in the organization who needs access to or is affected by a contract. It should also have exceptional security, and the ability to define roles, assign them to users, and override default permissions on a user-by-user basis.
- **Template repository**—The system should have a repository where legal and procurement teams can create standard terms, clauses, and contracts for common categories and services to streamline the procurement and contracting cycles. These templates can include statements of work, specifications, or standardized terms and conditions.

Supplier performance measurement and control are discussed in Chapter 6. E-sourcing tools can help monitor, measure, and provide visibility to supplier performance. Electronic scorecards are in real-time based on the data collected regarding delivery, quantity, quality, and others. Having real-time data can help the suppliers to detect problems early in the process. The e-sourcing tools can provide reports on variation, process control charts, and quality conformance.

Supplier performance systems also can perform a total cost comparison of suppliers' cost versus price. They provide real-time quality and other performance concerns. When it comes time for the suppliers' review, valid and reliable discussion on historical performance is accessible.

Total cost reporting are discussed throughout the book. The concept of total cost is important in decision making. Purchasers have to understand the cost of doing business with a particular supplier, in a particular region to make the right supplier decision, and help to mitigate supplier risk. These e-systems can again provide real-time data, making access to the data elements in the total cost model easier to assimilate. Total cost reporting can also help to monitor price variability and market pricing. Price is often the largest component of total cost and must be carefully managed.

Emerging Technology Issues and Capabilities

A number of improvements, issues, and capabilities in the technology support e-sourcing. These are discussed in the following sections.

Social Media and Mobile

One trend that will continue to impact the sourcing area is the use of professional networking software such as LinkedIn. LinkedIn is the world's largest professional networking group.[25] Professionals can use this network to connect with other supply managers, both within and outside of their organizations and share ideas. Both buyers and suppliers can leverage these types of sites to gain business advantage. Both buyers and suppliers can reach out to each other from/to previously untapped markets.

Sites such as LinkedIn also enable questions to be posted on the social network site; jobs can be posted; and other professional development activities are often available. Recommendations and referrals are also possible through its messaging system. There are also many special interest groups available on LinkedIn, including CSCMP, which currently has approximately 30,000 members.[26] There are many purchasing specific groups on LinkedIn; potential users can run a search and see what interests them.

In addition, a number of blogs (bloggers) are operational that focus on the area of sourcing management. For example, Forrestor, a blog for sourcing and vendor management professionals, helps purchasers understand particular issues facing business professionals. This blog "rolls-up" all the posts from analysts that serve sourcing and vendor management professionals.[27] These blogs are another opportunity for supply managers to access immediate information and ideas.

Twitter is an online social networking service and micro-blogging service that enables its users to send and read text-based messages of up to 140 characters.[28] Twitter has also started to have a presence in the area of purchasing and supply management; although, the use of tweets in B2B applications is relatively limited.

A research study was conducted in the brokerage industry for shipping. This industry has a number of rigid and old-fashioned rules of negotiations. Actually, contracts do not have to be written; instead they are "fixed" through a process of negotiation. In the past, all these negotiations were either face to face or via telephone. However, texting, email, and other new means of communication are starting to replace the telephone (or tele-fax).

New Global Sourcing Paradigm Through Cloud Strategy

The dynamics of the sourcing environment will change with cloud solutions. Cloud computing is Internet-based, and users share software and other information that is then provided on demand to computers and mobile devices. Dropbox is a common cloud application.[29] Common business applications are delivered by providers to online locations where the data is then hosted, and data can be accessed from anywhere.

Clouds can be private or public. A public cloud sells service to anyone on the Internet. A private cloud is a proprietary network or a data center that supplies hosted services to a limited number of people. The movement from hosting applications behind firewall systems to cloud computing has been significant. However, there are still many concerns about the risk associated with data floating in a cloud. Data security becomes a key concern.

Globalization and Big Data Analytics

This highly connected environment that is achieved through both mobile and social networking technologies is revolutionizing the role of purchasing. Organizations face exponential growth in the amount of data available. This is both an opportunity and a threat. Technologies that can harness the data and provide integration of information across multiple platforms can help to gain business advantage. However, data security threats are introduced, and companies have to look at ways to better manage and regulate the use of the data.

Users want 24/7 access to the e-sourcing systems from anywhere in the world. Users also want the information to be delivered to any mobile device including cell phones, iPads, and computers. Accessibility and availability of data in the software packages is an emerging trend that will continue to challenge providers. Another concern is using the systems to help global organizations understand best business practice, no matter the culture, business norms, infrastructures, and legal systems. The challenge is to collect and share these best organizational practices given some extremely challenging constraints.

Supply chain connectivity both internally and externally still remains a challenge. Connecting different entities across the supply chain creates efficiencies needed to gain a competitive advantage. Companies can create supplier portals to increase communication flows between buyers and suppliers. However, there can be significant push back from suppliers because of the requirements for sharing this information. Also, internally you have to keep in mind that sourcing has become inherently cross-functional, meaning that many other areas of the business want to access and use the e-sourcing tools.

However, with both internal and external integration, the e-sourcing tools must be user-friendly to streamline the processes and reduce unnecessary complexity.

Integration of technologies is a problem that organizations have faced since the first MRP/DRP systems were implemented. There are many add-on packages that perform a specific task, or legacy systems that are so incorporated into business processes, and can't be changed or eliminated. One technology trend is to focus on easily integrating these technologies into a more manageable package without increasing implementation time and cost.

Finally, tools that provide better decision support are necessary as the field continues to grow. The sourcing teams must be constantly alert to capabilities that can help enable business strategy. These may be in the form of technology or in the form of supplier innovation. Using technology to harness internal and external knowledge is key for strategic success.

Conclusion and Chapter Wrap-Up

The goal of this chapter is to gain a better understanding of the roles of technology in supporting supply management operations. Many of the key categories of supply management software and their capabilities are identified. Finally, emerging trends and technologies are assessed with social networking and cloud computing taking an ever-increasing role in B2B interactions.

Following are the key takeaways of this chapter:

- Discuss the role of technology in sourcing.
- Identify some of the different types of technology.
- Examine some of the trends that are impacting supply management operations, and understand how these trends may impact your organization.

Key Terms

- Supply chain operations
- Material Requirements Planning (MRP)
- Distribution Requirements Planning (DRP)
- Electronic Data Interchange (EDI)
- Enterprise Resource Planning (ERP)
- Cycle times

- Centralization
- Decentralization
- Arbitrage
- Commodity exchange
- Request for Quote (RFQ)
- Supplier Relationship Management (SRM)
- Total cost reporting
- Electronic scorecards

References

1. Lean-Manufacturing: Japan. “Material Requirements Planning.” Retrieved September 8, 2013, from http://www.lean-manufacturing-japan.com/scm-terminology/mrp-materials-requirements-planning.html.
2. Business Dictionary (2013). Material requirements planning (MRP-MRPI). Retrieved on September 18, 2013, from http://www.businessdictionary.com/definition/material-requirements-planning-MRP-MRP-I.html.
3. Business Dictionary (2013). Manufacturing Resource Planning (MRP-II). Retrieved September 18, 2013, from http://www.businessdictionary.com/definition/manufacturing-resource-planning-MRP-II.html.
4. Business Dictionary (2013). Distribution Requirement Planning (DRP-I). Retrieved on September 18, 2013, from http://www.businessdictionary.com/definition/distribution-requirement-planning-DRP-I.html.
5. Tech Terms. “EDI.” Retrieved on September 7, 2013, from http://www.techterms.com/definition/edi.
6. Reference for Business. Electronic Data Interchange. Retrieved on September 14, 2013, from http://www.referenceforbusiness.com/small/Di-Eq/Electronic-Data-Interchange.html.
7. Singleton, D. “Compare Enterprise Resource Planning Software.” Retrieved September 8, 2013, from http://erp.softwareadvice.com/.
8. NMA Technology. “What Is an ERP System?” Retrieved September 7, 2013, from http://www.nmatec.com/workwise/what.

9. Rajesh, K., excITingIP.com, "Advantages and Disadvantages of ERP Systems." Retrieved September 14, 2013, from http://www.excitingip.com/2010/advantages-disadvantages-of-erp-enterprise-resource-planning-systems/.
10. Ibid.
11. Dominick, C. "What is esourcing and how can it help you?" Retrieved September 7, 2013, from http://www.nextlevelpurchasing.com/articles/what-is-esourcing.html.
12. SAP (2013). Supplier Relationship Management Software. Retrieved September 18, 2013, from http://www54.sap.com/solution/lob/procurement/software/srm/index.html.
13. CRM Landmark (2013). CRM and software as a service glossary. Retrieved on September 18, 2013, from http://www.crmlandmark.com/resources_glossary.htm.
14. Fishman, C. (2006). *The Wal-Mart Effect: How the World's Most Powerful Company Really Works—and How It's Transforming the American Economy*. Penguin Books.
15. University of San Francisco, online education resources. "Walmart: Keys to Successful Supply Chain Management. Retrieved September 2, from http://www.usanfranonline.com/wal-mart-successful-supply-chain-management/.
16. Burt, David N., Dobler, Donald W., and Starling, Stephen L. (2003). *World Class Supply Management*, 7th Edition. McGraw-Hill: New York.
17. Flynn, Anna E. (2004). "Developing and Implementing E-Sourcing Strategy." CAPS Research Critical Issues Report, September.
18. Booz-Allen and Hamilton (2000). "E-sourcing: 21st Century Purchasing." Retrieved September 2, 2013, from http://www.boozallen.com/media/file/80568.pdf.
19. Booz-Allen, ibid.
20. Booz-Allen, ibid.
21. Monczka, et al. (2011), ibid.
22. General Mills, World-wide Sourcing, esourcing Supplier Guides. Retrieved September 2, 2013, from http://generalmills.com/~/media/Files/esourcing_training_guide.ashx.
23. Booz-Allen, ibid.
24. esourcing wiki. Contract management and compliance. Contract management 101. A Total Value Management Introduction. Retrieved on September 2, from http://esourcingwiki.com/index.php/Contract_Management_and_Compliance#Contract_Management_Defined.
25. LinkedIn. Accessed September 14, 2013, from http://www.linkedin.com/.

26. LinkedIn. Accessed September 14, 2013, from http://www.linkedin.com/groups?gid=100178&mostPopular=&trk=tyah&trkInfo=tas%3ACouncil%20of%20Supply%20C.
27. Forrester. Retrieved on September 14, 2013, from http://blogs.forrester.com/sourcing_and_vendor_mgt.
28. Twitter. Retrieved on September 14, 2013, from https://twitter.com/.
29. Dropbox. Retrieved on September 14, 2013, from https://www.dropbox.com/.

5

DEFINE THE REQUIREMENTS AND CHALLENGES OF SOURCING ON A GLOBAL BASIS

This chapter identifies the fundamental issues of global sourcing and the steps required to mitigate risk. It goes into more depth on some of the aspects of risk and points out many of the differences between global sourcing and domestic sourcing and some of the key decision-making criteria between the two.

Learning Objectives

After completing this chapter you should be able to:

- Understand the nuances between sourcing globally and sourcing domestically.
- Describe the key elements, risks, and benefits of global sourcing.
- Determine the qualitative and quantitative data needed in the geographic location decision over the total product and service life cycle.
- Further develop total cost of ownership models that consider potential supply disruption.
- Explain the geographic and sourcing risks inherent in global sourcing and how to potentially mitigate supply chain risk and disruptions.
- Describe the challenges and pitfalls in managing contract compliance and performance of global suppliers.

Introduction to Global Sourcing

The amount of global sourcing for the top 20 U.S. trading partners has increased significantly in the past decade. Recent global purchases of manufactured goods and services are shown in Table 5-1. Note that few countries have a positive trade balance. Also, in the United States the number of exports has continued to increase but the imports have remained relatively steady over the past 5 years. The phenomenal population growth in China has caused imports to climb significantly over the past decade.

Table 5-1 U.S./International transactions for 2007–2012[1]

	(Millions)	2007	2008	2009	2010	2011
Asia/Pacific	Export	549250	581845	512760	635072	714848
	Import	–976849	–978636	–822513	–982544	–1072104
	Total	–427599	–396791	–309753	–347472	–357256
Africa	Export	41570	49722	42704	49602	57779
	Import	–100291	–122702	–71496	–94559	–103957
	Total	–58721	–72980	–28792	–44957	–46178
Australia	Export	47226	52764	44509	54394	65178
	Import	–25219	–24327	–19696	–24217	–24113
	Total	22007	28437	24813	30177	41065
Europe	Export	899025	949564	749178	785415	878014
	Import	–933795	–962184	–716762	–788438	–882167
	Total	–34770	–12620	32416	–3023	–4153
S.&C.	Export	327333	377255	321335	395933	475911
America	import	–398887	–432767	–332644	–407817	–484365
	Total	–71554	–55512	–11309	–11884	–8454
USA	Export	1654561	1842682	1575037	1837577	2105045
	Import	–3083637	–3207834	–2427804	–2835620	–3182655
	Total	–1429076	–1365152	–852767	–998043	–1077610
Canada	Export	337891	363438	282460	350074	401269
	Import	–295106	–331841	–356441	–369648	–390437
	Total	42785	31597	–73981	–19574	10832

	(Millions)	2007	2008	2009	2010	2011
China	Export	85846	96358	95488	126649	143051
	Import	−216767	−272295	−331378	−378414	−400564
	Total	−130921	−175937	−235890	−251765	−257513
Brazil	Export	46351	59598	50931	67622	81962
	Import	−35309	−42489	−29591	−33424	−43358
	Total	11042	17109	21340	34198	38604
India	Export	27631	32014	30208	34725	39241
	Import	−19223	−25261	−30864	−35891	−40006
	Total	8408	6753	−656	−1166	−765
Japan	Export	131126	135821	118288	129831	138819
	Import	−203747	−175811	−175470	−170514	−198104
	Total	−72621	−39990	−57182	−40683	−59285
Mexico	Export	177131	191966	162529	200849	239429
	Import	−247978	−254745	−207126	−260079	−295235
	Total	−70847	−62779	−44597	−59230	−55806
EU	Export	780382	805999	639614	660643	726049
	Import	−816174	−809288	−608956	−666235	−736328
	Total	−35792	−3289	30658	−5592	−10279
Middle East	Export	76192	88776	73109	81504	100235
	Import	−111253−	147121	−89326	−103151	−132982
	Total	−35061	−58345	−16217	−21647	−32747

For sourcing, the purchases from some areas, such as China, have increased significantly over time, whereas the exports to China have not increased at the same rate (Table 5-2). Other locations, such as Mexico, have fallen out of favor in the last decade. Although, it is one of the countries often considered as organizations think about reshoring or near-shoring. Other interesting areas of the world are India and Africa. Just looking at the variability over time causes a purchaser to include an in-depth country analysis on each of these areas. Even countries with a historically greater risk associated with trade have seen significant increases in imports to the United States and exports from the United States. For example, trade with Saudi Arabia has almost quadrupled in the past decade. Purchasing plays a key role in determining what locations make sense for its suppliers.

Table 5-2 U.S. Trade with India, China, Africa, and Saudi Arabia (Exports and Imports)

U.S. trade in goods with India						
Exports	$14,968.80	$17,682.10	$16,441.40	$19,248.90	$21,503.50	$22,105.50
Imports	$24,073.30	$25,704.40	$21,166.00	$29,532.90	$36,155.50	$40,514.10
Balance	49,104.50	48,022.30	44,724.60	410.284.00	414,652.00	418,408.60
U.S. trade in goods with China						
Exports	$62,936.90	$69,732.80	$69,496.70	$91,911.10	$103,986.50	$110,483.60
Imports	$321,442.90	$337,772.60	$296,373.90	$364,952.60	$399,378.90	$425,578.90
Balance	4258,506.00	4268,039.80	4226.877.20	4273,041.50	4295,392.40	4315.095.30
U.S. trade in goods with Africa						
Exports	$23,425.80	$28,392.70	$24,329.50	$28,339.90	$32,841.30	$32,737.50
Imports	$92,013.10	$113,495.60	$62,403.60	$85,008.10	$93,009.10	$66,817.10
Balance	468,587.30	485,102.90	438,074.10	456,668.20	460,167.80	434,079.60
U.S. trade in goods with Saudi Arabia						
Exports	$10,395.90	$12,484.20	$10,792.20	$11,506.20	$13,826.60	$17,972.00
Imports	$35,626.00	$54,747.40	$22,053.10	$31,412.80	$47,476.30	$55,666.30
Balance	425,230.10	442,263.20	411,260.90	419,806.60	433,649.70	437.694.90

Chapters 1 and 2 both discuss some strategies for deciding on the appropriate supplier location, and how to identify and then qualify suppliers. One goal of purchasing has always been to keep costs down. Targeting the *low labor-cost regions* of the world is a tactic that may help organizations remain competitive and survive in the competitive marketplace. However, low labor costs or emerging areas could potentially add costs in other areas, such as transportation.[2] Specific geographies for suppliers are often targeted to maintain a competitive edge, introduce competition in pricing, gain access to potentially new customer markets, or in reaction to the choices of other firms.

Global sourcing opens up a myriad of opportunities for supply managers. However, the potential rewards of global sourcing go hand-in-hand with risks and challenges stemming from rapid global growth, inadequate controls around current processes, and the difficulty of managing the technological interface. [3] So, supply managers must be cognizant of opportunities that are available in other regions of the world but can carefully assess these opportunities using the tools, methods, and processes discussed in earlier chapters.

Using the global supply base and developing supply relationships that deliver improved value to end customers in terms of cost, quality, delivery, and performance helps supply managers plays a strategic role in the organization and achieve a competitive advantage. As mentioned in Chapter 4, the Internet and other technology improvements have helped to accelerate the trend to global sourcing, making it easier for source selection and reducing communication problems. The increased speed of transportation and communication has helped decrease the size of global landscape.

Cloud services and mobile networking in B2B interactions are expected to grow exponentially in the upcoming years making instant data access available in areas all around the world.[4] The use of social media, blogs, and discussion forums is expected to have an impact on the sourcing decision-making process. Many regions of the world target certain industries, and suppliers become specialized. Purchasers can use social media, blogs, and discussion forums to keep up on these types of trends.

Managing global supply networks presents a number of challenges in areas such as source identification and evaluation, international logistics and country infrastructures, misaligned communication and information systems, changing cost structures, and currency rates. Global sourcing also introduces a number of risk factors not necessarily present in domestic sourcing.

Global Sourcing

The underlying reason for using an offshore supplier is that better value is perceived to be available from that source than from a domestic one. A supplier's ability to be competitive is influenced by the overall competitiveness of the country. As mentioned in earlier chapters a thorough analysis of different options is necessary. Embarking on a global sourcing strategy requires a level of country knowledge and analysis far beyond that for a domestic supplier.

Advantages of Global Sourcing

The specific factors that make the international buy look attractive vary from country to country and also change over time. A stronger U.S. dollar makes the price of goods offshore more attractive. Some of the primary reasons to select an offshore supplier as the preferred source are discussed next.

Case in Point—Changes in Perceived Location Decisions

Previously, the perception of Japan's products was that it had extremely poor quality. Many people described it as "junk" and said they would never buy products made in Japan. Japan made a big push to improve its perception and hired quality gurus such as Deming and Juran to help improve quality levels and to take advantage of the growth in international markets. Within a relatively short period of time, the products in Japan were seen as superior to those available in other parts of the world. Soon, the United States lost most of its market share of technologically sophisticated products: cell phones, computers, televisions, and so on. The Japanese figured out a way to improve their products by improving quality and gained a significant competitive advantage. Today, most consumers have some Japanese products in their possession.

Unavailability of Items Domestically

Availability of materials is one of the oldest reasons for international trade; think of Christopher Columbus, for example! Items such as cocoa, coffee, spices, fruits, chrome, and others are only available from certain countries. There are many other resources, such as skilled labor and management that can be found only in certain areas of the world. For example, Apple recently moved production of one of its products back to the United States. The move was radically different than its current operating strategy. However, it was only a small part of the business, mainly because the United States does not have enough trained engineers to support its needed levels of production, so it targets areas of the world where there is a large pool of highly educated engineers that can help support its production needs.

Price and Total Cost Advantage

The ability for an offshore supplier to deliver products at a lower overall cost than domestic suppliers is a key reason to buy globally. There are many reasons for the costing advantage in certain locations. Supply managers have to be careful that they are not just looking at "price" but instead are looking at the "total cost" of business. They must be aware of the additional costs added due to regulation, transportation, taxes, and other areas. Quality costs can also be significant in offshore locations. Specific cost elements will be discussed later in this chapter, in general some of the cost elements that drive a lower total cost include

- Labor costs that are substantially lower
- Exchange rates that favor global sourcing

- Differing and potentially more efficient equipment, technology, and processes
- Supplier pricing models, especially that support regional production of specific types of goods

Government Pressures and Trade Regulations

Trade incentives or restrictions may influence decisions about source location. Quotas, reciprocal trading, bilateral trading agreements, offsets, and others can determine a specific sourcing location. Pressures and regulations, such as currency, have a tendency to fluctuate over time depending on the political relationship of the United States with a given country. Many times these regulations focus specifically on environmental or land use issues. For example, a solid wood furniture manufacturer ended up requiring fumigation of all containers coming from China to get rid of beetles that were not indigenous to the United States. The fumigation effort, the certification effort, and all the other associated paperwork added significant costs to doing business with the supplier.

Quality

In some cases, quality may be better or just as consistent from offshore suppliers compared to domestic suppliers. Many offshore locations hold their workforce responsible for "doing it right the first time" and quality systems and management of those systems are world-class.

The concern with quality is that if there is a problem what costs might be incurred, what is the process for returning the item, and what is the obligation of the supplier to provide spare parts for repairs? The costs associated with returning an item with an extending shipping channel is exorbitant; often it is better to repair these items on site or to simply downgrade or dispose of the item and charge back the supplier.

Spare parts and hardware are often challenging and costly to receive. Is it possible to maintain a stock of these items? What is the associated cost of doing that? How much additional labor costs is associated with part replacement or repair? These are only some of the issues that make offshoring more challenging than domestic sourcing and drive up the costs of doing business with a particular supplier. Maintaining the appropriate level of quality is difficult and companies have tried many different ways including co-location of its own employees to the supplier's site. [5]

Faster Delivery and Continuity of Supply

Even with the longer transportation pipeline, it is sometimes possible to maintain the continuity of supply and thereby achieve faster delivery. Suppliers in global locations will often hold additional inventory to ensure that products will arrive when needed. Many developing countries have also invested heavily in infrastructural improvements to facilitate the movement of products. Also, there are free-trade zones established to facilitate easier movement of goods.

Marketing Tool

Many of these developing countries also have a large emerging customer base. Through global sourcing an organization may gain access to that market share. It makes sense to enter a market to try and gain access to the developing customer base. The suppliers' employees start talking about your organization and may feel a sense of obligation to buy the products.

Competitive Clout

Offshore suppliers may help to introduce competition (and therefore reduce pricing) to existing suppliers. The total cost of doing business with an offshore supplier may benefit the buyer in negotiations with other suppliers in other regions of the world. For example, developing a total cost model for a supplier in China with a low price, but high risk, can open the door to negotiate with a supplier in the United States for a price reduction.

Disadvantages of Global Sourcing

There are also many potential problem areas in sourcing globally. Purchasers have to consider the total cost and not fall into the trap of focusing on lowest price. Cross-functional teams are necessary to ensure that the offshore supplier can perform to the level of expectation. A thorough analysis of the supplier using the tools discussed in earlier chapters is needed. Other potential problem areas are discussed next.

Source Location and Evaluation

If the associated level of spend and risks are high, a site visit is necessary. Of course, this adds to the cost of doing business with the potential supplier, as will travel that might be needed for performance. During the site visit, consider how often a visit might be necessary; this too becomes part of the total cost.

Evaluate global suppliers as you would domestic suppliers; don't make any assumptions about capabilities. Performing a location specific risk assessment is also necessary. There are many examples in the current public press where suppliers were not appropriately screened discussed in earlier chapters—Bangladesh, FoxConn, and others have had serious violations of workers' rights and also of safety in the workplace.

Lead Time and Delivery

The extended pipeline of products makes it more challenging to coordinate materials. Purchasers have to work closely with the suppliers, freight forwarders, and transportation companies that need to know where the products are, and the expected time of arrival. Other unplanned delays such as customs, duties, regulations, and port operations may create problems that were not anticipated. Changes to production plans and expediting are significantly more challenging. In some cases using the 45' high-cube containers is more efficient. However, in some countries the infrastructure does not support the movement of these larger containers. For example, in Vietnam, a worker was employed to push electrical lines up with a stick, so the high cube trucks preferred in the furniture industry could move easily through the country.[6]

Political, Labor, and Security Problems

Risk management has been discussed throughout this book and is something to be concerned about. Some countries have other challenges that need to be considered before sourcing from a supplier located there. There is a heightened risk of supply disruption from terrorist acts, counterfeit goods, and unsafe products. Risk management strategies and contingencies are important in a global economy.

The early 2000s brought back the idea of "pirates" on the high seas. Shipping through some locations was both challenging and dangerous. Crews were trained to manage an invasion of the ships, which often resulted in deaths and injuries to save cargo. There are many stories where the insurgence of people in a particular economy caused significant disruption to the supply chain.

Hidden Costs

As mentioned numerous times, a buyer must be aware of potential hidden costs that impact the total cost of doing business with a supplier. There is also the potential for significant variability in costs causing the total cost to fluctuate. For example, the cost of transportation has increased significantly over time due to increasing fuel costs (see Figure 5-1). Changes in customs documentation often create delays and increase costs; currency fluctuations may have a material impact on purchases. Many other potential hidden costs exist as well. Purchasers and the purchasing team must do a thorough and careful analysis of the specific country, the specific commodity, and the specific supplier. For example, Figure 5-1 shows how diesel fuel prices vary over time, which impacts the total cost.

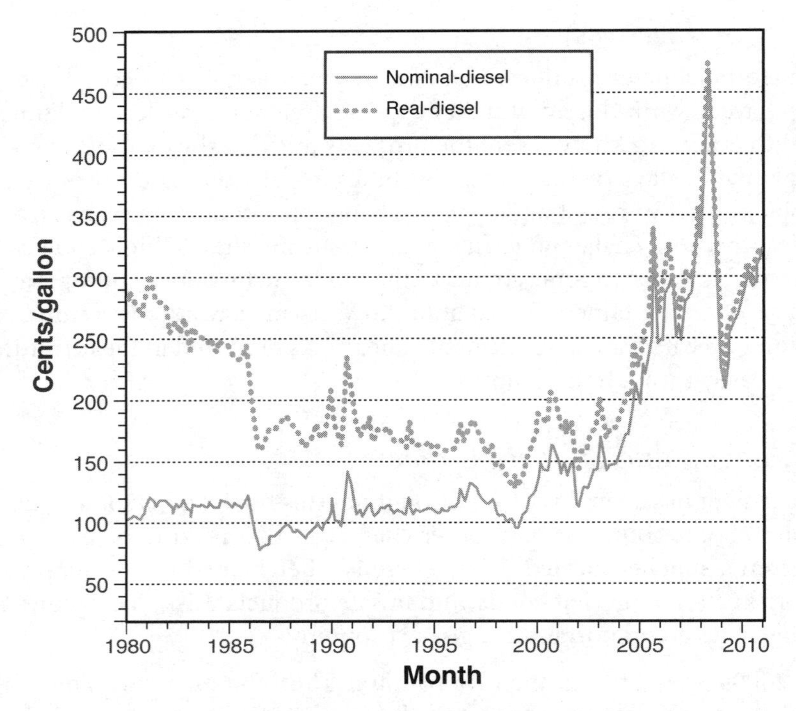

Figure 5-1 Monthly diesel fuel prices[7]

Quality, Warranties, Claims, and Replacement Parts

A fluctuation in quality also can have significant impact on the total cost. If there is a problem and the parts do not meet the quality standard, it is challenging to return the items over such long distances as discussed earlier. Companies must clearly discuss the options up front: who is going to pay for repair, downgrading, or disposal? These must be discussed during the contract negotiation and all policies must be explicitly stated.

Language, Communication, and Cultural and Social Customs

Words mean different things in different countries and the presentation is critical. The problems intensify when using electronic communications (email and texting) because of the pace of communication. There are time zone issues and problems with the communication networks. Differences in language must be considered in the overarching country analysis. The use of interpreters may add an additional cost of doing business. Not having an interpreter may add even more costs if the contracts are not appropriately understood by both parties. Educating buyers on country specific nuances may also be costly to the buying organization. Cultural and social customs vary significantly across geographic borders. Be certain that these customs are understood before finalizing the selection decision.

Case in Point—Cultural Differences

Doing business in a global setting is challenging because a purchaser must understand "how" to do business in a different cultural setting. Understanding the expectations during meetings, negotiations, meals, and so on can set you apart from the competition.

The importance of punctuality changes across borders. Some countries, such as Germany, expect every meeting to start and end on time. On the other hand, countries such as Brazil likely start late and end late. Purchasers can't get frustrated when there is a more flexible attitude to time.[8] In Brazil, people consider the relationship far more important than time-related issues.

Women in Saudi Arabia play virtually no role in business. It is difficult for women business travelers to achieve a great deal in the country. The delegation sent to Saudi needs careful consideration. Typical Western concerns on selection to the delegation based on achievement, track record, or technical capabilities may not be appropriate.[9]

Respect and relationship is of utmost importance in China. One must always respect seniority and never openly disagree with people because they could "lose face." Business cards in China should be formally exchanged at the beginning of meetings. Unlike the United States and other regions, the business card represents the man. Age, seniority, and educational background are critical to recognize.

There are many differences across locations because of culture. There are some that you don't even think about that could have disastrous impact on your negotiations or even your business relationship. Take the time to *thoroughly* understand the culture; there are many websites, consultants, and potentially other employees that can provide information that you may need.

Ethics and Social and Environmental Responsibility

There are laws and regulations in place to monitor the supplier's behavior and buyer's behavior. For example, the Foreign Corrupt Practices Act was passed in 1977, which prevents U.S. firms from providing or offering payments to officials of foreign governments to obtain special advantages. More companies also report on their corporate social responsibility. A number of formal monitoring and measurements are required for global trade and supply chain practices. Supply chain visibility and accountability are increasingly important in a global environment. Many companies have also implemented a Code of Conduct for the supply base that requires responsible actions of suppliers in challenging areas. Because of the many well-publicized issues in industries such as electronics, a specific code of conduct has been instituted. For example, Table 5-3 shows the broad categories of the EICC.

Table 5-3: EICC Code of Conduct, Major Sections, and Supplier Responsibilities[10]

EICC Code of Conduct
The Code of Conduct provides guidelines for performance and compliance with critical CSR policies. EICC provides tools to audit compliance with the code and helps companies report progress. The Code is made up of five sections. Sections A, B, and C outline standards for Labor, Health and Safety, and the Environment, respectively. Section D adds standards related to business ethics; Section E outlines the elements of an acceptable system to manage conformity to this Code. Supplier Responsibilities related to EICC: ▪ Phase One: Organizational Assessment ▪ Phase Two: Self-Assessment & Training ▪ Phase Three: Validated Audit

Geographic Location Decision—Dealing Directly with Offshore Suppliers

Dealing directly with global suppliers results in the lowest purchase price because it eliminates the mark-ups associated with intermediaries (adding another potential cost element to the total cost model). However, this model also requires investment in travel, communications, logistics, and interpretation of the various cost elements. The decision to source globally and deal directly with suppliers should be undertaken only after a careful assessment of available opportunities both globally and domestically. Buyers should learn to anticipate problems.

Before proceeding with the supplier selection process, two important issues must be addressed: country and regional stability and the potential supplier's financial condition. You can purchase information about a particular region and a particular supplier from organizations such as Dun & Bradstreet. However, many other questions need answers before committing to a particular global supplier. These include but are not limited to the following:

- Are there political or monetary stability issues to consider?
- What is the required documentation for transactions?
- How well developed is the transportation and distribution infrastructure?
- Are there particular religious customs or holidays that will impact supply?
- What are the current quality standards?
- Will trade and product liability policies impact business?
- Are there existing regulations that might restrict the sale of the product or service?
- Will there be agents that can help to facilitate the contract?

D&B Country Risk Indicator[11]

D&B's DB risk indicator provides a comparative, cross-border assessment of the risk of doing business in a country. Essentially, the indicator seeks to encapsulate the risk that countrywide factors pose to the predictability of export payments and investment returns over a time horizon of 2 years.

The DB risk indicator is composed of a composite index of four overarching country risk categories.

Political Risk

The internal and external security situation, policy competency and consistency, and other such factors determine whether a country fosters an enabling business environment.

Commercial Risk

The sanctity of contract, judicial competence, regulatory transparency, degree of systemic corruption, and other such factors determine whether the business environment facilitates the conduct of commercial transactions.

Macroeconomic Risk

The inflation rate, fiscal deficit, money supply growth and all such macroeconomic factors determine whether a country can deliver sustainable economic growth and a commensurate expansion in business opportunities.

External Risk

The current account balance, capital flows, foreign exchange reserves, size of external debt and all such factors determine whether a country can generate enough foreign exchange to meet its trade and foreign investment liabilities.

Cultural Preparation

As mentioned in earlier chapters, negotiating with suppliers located in other regions of the world has some special nuances. The success of negotiations with a global supplier is influenced largely by the negotiator's ability to understand the needs and ways of thinking and acting of the representatives of the global firm. What is considered ethical in one country is not ethical in another. The intention of filling commitments, the implications of gift giving, and even the legal systems vary considerably. In addition to the conventional preparation for any negotiation (discussed in earlier chapters), it is essential to conduct an extensive study of the culture. Cultural preparation is specific to the country in which purchasers plans to conduct business.

Technical and Commercial Analysis

Before dealing with identified potential suppliers, additional items must be carefully prepared before discussions:

- **Specifications and drawings**—Make sure that all specifications are clear and understandable; this includes engineering change management orders. Many organizations have a difficult time introducing and implementing changes to specifications. The engineering department of a company was consistently contacting offshore suppliers with changes. The changes were sometimes miniscule but others relatively significant. The supplier ended up with so much obsolete product in its facilities that it had no choice but to increase the price.
- **Samples or photos**—Prepare samples and photos to help in communicating the details of the item to purchase. The old adage is that a picture is worth 1,000 words. In this case it is because the words mean different things when said in a different language.
- **Quality requirements**—Carefully prepare the appropriate quality requirements. Currently, there is a tendency to "over-specify" quality needs with global suppliers. This practice is not in the best interests of the supplier or of the buyer because ultimately it causes increased costs. Be realistic. Do you need 98.5 percent quality specifications, or for this particular commodity will 85 percent work?
- **Scheduling requirements**—Understand the specific timing and needs for these components or products. Production plans, forecasts, and other scheduling information should be considered. Also, consider the lead-time involved for production and transportation. Discuss how long in the future the schedule is "fixed" and what happens when an emergency order comes in.
- **Offshore versus domestic production**—What are the annual requirements for the item, and how much of it can be sourced globally versus domestically? This

involves discussion with distribution center managers, inventory control managers, and others to determine the best policy and the least associated risk.

- **Packaging**—Because of the increased transportation pipeline, it may be necessary to have additional packaging to protect the products from damage in transit. Some of the roads can be much rougher in other countries and cause more in-transit damage. Ocean containers can be handled multiple times, again causing damage. Don't over-specify, but in this area don't under-specify either.
- **Pricing**—Think about the pricing objective before entering into discussions with suppliers. This should be a part of your planning for negotiation. A manufacturer was offshoring production of an item for the first time. This manufacturer collected data on some of the elements of total cost and then asked the supplier what the price would be on each item that was involved. The supplier's response was: What do you want it to be? Pretty much makes you wonder about the cost structures and the cost of materials.

Currency and Payment Issues

Exchange rates can have a significant impact on the total cost. There are at least two potential situations in which the absence of a fixed exchange rate can create a problem.

If a contract calls for payment in a foreign currency (that is, Euros) and the exchange rate moves against the U.S. dollar during performance of the contract, the following situation could occur.

Example: The contract was for $100,000 Euros, and the exchange rate at the time of contracting was $1 USD = .75 Euros. The cost in U.S. dollars to the buyer would be

Euro 100,000 = **$133,333.33**

Euro .75/$

If the U.S. dollar strengthens to $1 USD = .85 Euros, the situation changes. In this situation the U.S. buyer would benefit but the supplier would be no better off.

Euro 100,000 = **$117,647.06**

Euro .85/$

If the U.S. weakens to $1 USD = .65 Euros, the situation changes for the buyer. However, the supplier is no better or worse off.

Euro 100,000 = **$153,846.15**

Euro .65/$

As you can see, exchange rates can materially impact the contract, and buyers must be cognizant of this issue.

Total Cost of Ownership

The total cost of ownership has already been mentioned many times throughout this chapter and other chapters. The most important thing is that all the cost elements are identified that can materially impact doing business with a particular supplier in a particular location. The total cost of ownership analysis should include a comparison of the potential cost elements between domestic and global suppliers. Table 5-4 is presented for comparison. This is a total cost comparison of four different countries for small engines for lawn mowers. Part of this issue is that the company wants to manage and mitigate risk especially of supply disruption and environmental risk.

Table 5-4 Total Cost with an Environmental Twist

	China (Container)	Eastern Europe (Container)	Western Europe (Container)	U.S.(Each)
Unit Price	$42,000.000	$60,000.000	$85,000.000	$120.000
Packaging	$2,000.000	$2,500.000	$1,750.000	$1.000
Tooling	$100.000	$75.000	$125.000	$0.025
Inland Transportation	$200.000	$750.000	$200.000	$5.200
Ocean Transportation	$3,000.000	$4,500.000	$3,500.000	
Freight Fwder Fee	$200.000	$400.000	$100.000	
Insurance	$420.000	$600.000	$850.000	
Brokerage Fee	$400.000	$400.000	$400.000	
Travel	$250.000	$250.000	$250.000	
Administrative	$150.000	$200.000	$75.000	
U.S. Port Handling Charge	$1,500.000	$1,500.000	$1,500.000	
Duty	$2,100.000	$3,000.000	$4,250.000	
Brkerage 2	$500.000	$500.000	$500.000	
Port to Avondale	$1,500.000	$1,500.000	$1,500.000	
Cost of Capital	$840.000	$1,200.000	$1,700.000	
Warehouse Costs	$1,600.000	$1,600.000	$1,600.000	
Defect Rate	$2,100.000	$1,800.000	$850.000	

	China (Container)	Eastern Europe (Container)	Western Europe (Container)	U.S.(Each)
Environmental Surcharge			$500.000	
Environmental Surcharge			$750.000	
Price for Container	$58,860.000	$80,775.000	$105,400.000	
Price Each	$58.860	$80.775	$105.400	$126.230

The buyer is looking for a supplier and is trying to decide on the most important criteria for making a selection. Each of these countries has significantly different regulations and processes for manufacturing. Also, the initial price paid to the supplier varies significantly. A purchaser has to develop these types of models without neglecting any of the significant cost elements and then make a determination of the right source for supply. Some questions to consider in developing and assessing a total cost model follow:

- Have all relevant cost elements been included?
- Are there things to consider that might change and materially impact the total cost (that is, quality or transportation)?
- Are there other risk factors to consider with each of the locations?
- Is there a possibility to negotiate on price with any of the potential suppliers?
- Are there ways to reduce the cost elements that might change the total cost?
- What is in the best interest of the buying company—socially, economically, and environmentally?
- What makes the most business sense?
- How should you proceed?

A total cost model can be a powerful tool for international decision making. However, a total cost model is only a snapshot in time. It represents only the cost of doing business on the day that it was developed. The market and other aspects fluctuate. Understanding how this variability may impact the business relationship with a particular supplier is often challenging.

Supply Chain Risk and Disruption Management

Managing the global supply chain is challenging. However, the goal of a purchaser is to ensure an uninterrupted flow of supply in an efficient and effective manner. The interruptions are often difficult to predict, but recognizing the different types of risk may help to better manage a potential disruption. Some of the potential risks have been discussed in previous sections, such as currency risk and the risk of maintaining fewer suppliers or over-rationalizing the supply base. However, there are many other areas of risk to consider as you perform a country and supplier analysis in the global environment. In addition to a total cost of ownership model, many organizations have developed a risk assessment model that rates and weights the different categories of risk. The following list contains a number of risk factors that might be included in the model. These can also be weighted like the scorecards discussed earlier. A comprehensive risk assessment performed on a regular basis can help purchasers to manage all the potential risk associated with global sourcing.

This list is not meant to be all-encompassing however; supply managers should use this as a starting point in assessing and therefore mitigating risk. Risk has been discussed throughout this book because one of the primary goals of a purchaser is to minimize risk.

Thinking back to some of the tools used in other chapters, Porter's 5-Forces Model, for example, can help you think about many of the categories of risk. Performing a spend analysis, finding the appropriate supplier tool, and many other parts of global sourcing simply magnify the risk. Following are some areas that you need to assess in a global sourcing market.

Market Risk

- Number of buyers competing for the same goods and services
- Reduction of product life cycles
- Emerging technology
- Trade secret and intellectual property protection
- Supplier competition and power
- Brand and reputation risk

Political risk

- Country stability
- Regional stability
- Political and governmental stability

Legal and differences in contract law

- Terrorism and levels of corruption
- Military and civil disturbances
- Intellectual property rights

Financial risk

- Supplier's financial stability and future viability
- Investments in technology, capacity, and supplier capabilities
- Currency exchange
- Contract compliance
- Duties
- Tariffs
- Taxes
- Inventory carrying costs

Operational or sourcing risk

- Length of supply chain

Technology risk

- Quality
- Material risk
- Social, ethical, and environmental risk
- Supplier labor force disruption
- Risk of natural disaster
- Risk of noncompliance
- Communication risk

In an environment that includes global sourcing, supply management needs to work closely with its suppliers to drive cost-savings, reduce supply disruptions, minimize risk, and deliver innovation and value to customers. It is much more effective to deal with and plan for risk than it is to try to recover from a disruption or disaster. There are ways to help mitigate risk such as multiple country sourcing, the use of knowledgeable intermediaries, and the use of third parties to gather and report on relevant information (such as Dunn and Bradstreet). Scenario analysis and brainstorming sessions to identify potential risk and consider the likelihood of those risks occurring can help the supply manager better manage risk issues.

Supplier Contract Compliance

There are many challenges and pitfalls in managing contract compliance and performance of suppliers with global sources of supply. Companies tend to make risky deals—and often make promises that they cannot keep. Contracts are signed without fully reading or understanding the multiple clauses contained within the body of the contract. Part of this is due to the idea that if there are suppliers located in a place where they are not easy to visit, add every restriction into the contract imaginable. As discussed earlier, one organization that was contracting for call center services showed up at the supplier with a 200-page contract!

Purchasers often make the assumption that the terms of one market are acceptable in all markets and do not realize the cultural and legal differences. In a global environment, purchasers have to consider ongoing commitment to the suppliers and relationship management. Often purchasers allow the contractual terms to supersede the relationship aspects. Some organizations focus on "preventative contracting," which involves spending more time in contract development. There also has to be some flexibility written in to the contract, maybe as it relates to market pricing volatility or changes to a production schedule. The volatile markets such as steel have these pricing volatility changes included. Because fuel prices are so variable, anything tangentially or directly related, such as transportation, should have volatility clauses included in the contract.

The common understanding in contract law is that it usually costs less to avoid getting into trouble than it does to pay for getting out of trouble. Spend the time up front to discuss expectations, define language and terminology, and ensure communication of potential contingencies in the contract. Purchasers must understand the underlying legal aspects of business transactions. They also must manage contracts and agreement on a day-to-day basis, which may require a different set of skills.

Conclusion and Chapter Wrap-Up

The intent of this chapter is to provide an overview of global sourcing and to point out some difference between global and domestic sourcing. The major differences revolve around risk, costs, and contract management. One major takeaway from this chapter is to make the purchaser aware of the challenges of global sourcing. The second major takeaway is in identifying and therefore managing the costs. There is more discussion provided on the idea of risk mitigation and country analysis.

Other key topics include

- An understanding of some of the nuances between sourcing globally and sourcing domestically

- An ability to describe and define some of the key elements, risks, and benefits of global sourcing
- The idea of including both qualitative and quantitative data needed in an assessment of the location decision over the total product and service life cycle
- Further capabilities in developing total cost of ownership models that consider potential supply disruption
- An understanding of the geographic and sourcing risks inherent in global sourcing and how to potentially mitigate supply chain risk and disruptions
- Identification and awareness of the challenges and pitfalls in managing contract compliance and performance of global suppliers

Key Terms

- Global sourcing
- Cost elements
- Technical analysis
- Commercial analysis
- Exchange rate
- Total cost of ownership
- Cultural preparation
- Supply chain risk
- Disruption management
- Supplier contract compliance

References

1. U.S. Department of Commerce, Bureau of Labor Statistics, International Transactions, www.bea.gov.
2. Tate, Ellram, Schonherr, Petersen (Forthcoming). Manufacturing location decisions. *Business Horizons.*
3. PWC (2010). "Why Global Sourcing? Why now? Creating Competitive Advantage in Today's Volatile Marketplace." Advisory Services. Retrieved on September 3, 2013, from http://www.pwc.com/en_us/us/increasing-it-effectiveness/assets/global-sourcing.pdf.

4. Mukherji, P. (2013). "Trends likely to Impact Global Sourcing in 2013." Globalization. Retrieved September 3, 2013, from http://www.globalizationx.com/global-sourcing-trends-in-2013/.
5. Tate, et al. *Journal of Service Research.*
6. Ellram, Tate, Feitzinger (2013). Factor Market Rivalry. *Journal of Supply Chain Management.*
7. Ibendahl, G. (2012). "Seasonality of Diesal Fuel Prices." *Journal of ASFRMA.* Retrieved on September 10, 2013, from http://www.asfmra.org/wp-content/uploads/2012/06/360_Ibendahl.pdf.
8. World Business Culture (2013). Business Culture on the World Stage. Retrieved September 19, 2013, from http://www.worldbusinessculture.com/Business-Meetings-in-Brazil.html.
9. World Business Culture (2013). Women in Business in Saudi Arabia. Retrieved September 19, 2013, from http://www.worldbusinessculture.com/Women-in-Business-in-Saudi-Arabia.html.
10. Electronic Industry Citizenship Coalition (EICC) (2013). Code of Conduct. Retrieved on September 19, 2013, from http://www.eicc.info/documents/EICCCodeofConductEnglish.pdf.
11. Dunn and Bradstreet (2013). Country Risk Indicator: Risk Services. Retrieved September 27, 2013, from http://www.dnbcountryrisk.com/.

6

ASSESSING THE INTERNAL AND EXTERNAL PERFORMANCE OF SUPPLY MANAGEMENT OPERATIONS

The purpose of this chapter is to gain an understanding of the internal and external performance assessment of supply chain and supply management operations. You learn about the consideration of the role of measurement, different types of measures, and the various trade-offs associated with performance metrics. An understanding of how to develop the right metrics to drive the right behavior is presented to show the value of managing supply processes with data that helps to achieve better and quantifiable results.

Learning Objectives

After completing this chapter, you should be able to:

- Understand the role of measurement to establish quantitative purchasing goals and key performance indicators.
- Develop integrative performance metrics that measure purchasing performance on other areas of the organization.
- Provide a review of the different categories of supply management performance measurement and key performance indicators.
- Describe various performance monitoring and improvement options for purchasing such as scorecards, dashboards, and exception management techniques.

Introduction to Assessment in Supply Management

Purchasing is a major contributor to organizational strategy and success. There should be alignment between organizational strategy and functional strategy, and purchasers directly contribute to organizational strategy by implementing specific objectives and effectively managing the supply base to deliver value to both internal and external customers. This chapter focuses on how to capture, measure, and communicate the contribution of supply management.

Peter Drucker was famous for saying, "You can't manage something if you are unable to effectively measure it",[1] and if you can't measure it, you can't improve it. [2] Applied to the complex world of purchasing and supply management, this means that all the processes, including strategic sourcing, negotiations, cost reductions, and customer and supplier management must be evaluated against goals, benchmarks, or on another designated scale.[3] By measuring the contribution to value and assessing the results, appropriate action can be taken to either incentivize, reward, or develop employees.

You must ensure that the contribution made impacts the organization's top and bottom line and also helps to achieve the strategic goals of the firm. Often, the establishment of performance measures drives the wrong behavior and shifts the focus away from strategic alignment and instead to functional or personal achievements. The challenge lies beyond what to measure, how to implement, and how to ensure compliance. You need a way to interpret the measures and act upon the results to drive behavior that facilitates benefits to the organization.

Role of Measurement

In today's competitive environment, the success of any company depends on the capabilities of the supply base and the effectiveness of the purchasing organization in harnessing those capabilities and managing the supply base. As discussed previously, more of companies' nonstrategic activities and much of the manufacturing is outsourced making organizations dependent upon their supply base for critical inputs into their offerings. The supply base also represents an opportunity to gather information on new technology and other innovation. The percentage of revenue spent on purchased goods and services continues to increase, as shown in earlier chapters.

Given the importance of the supply base to competitive success, it is vitally important that companies have in place metrics that measure the performance of the purchasing function and the performance of the supply base. Strategic metrics and operational metrics are required to effectively manage the performance and contribution of the department.

Measures and goals help to focus attention on the activities that are most important to achieving the corporate goals. Measures and goals must also align with various business units and other organizational functions. These joint metrics are necessary to dictate priorities, provide motivation, and track progress. Much of what the purchasing area does is performed in cross-functional teams. Good measures can help focus the attention of the team across business units and geographies. The purchasing organization must communicate clearly, reliably, and validly its contribution to the overall success of the organization.

Trade-Off Analysis

One of the key issues with performance measures is that they tend to promote behavior that is in the interest of the individual rather than the group. Also, there can be misalignment of the measures between corporate strategy and functional objectives. The performance measures must be aligned vertically with corporate goals and horizontally with the business units and other functions. The goal is to drive the appropriate behavior. Purchasing has long been accused of having a pricing orientation with complete disregard for the goals of other units. Purchasing would buy large volumes to get the price down and not worry what happened when the component ended up in inventory, or ultimately as obsolete. Both inventory and SKU proliferation occurred when purchasing did not act in the best interests of the organization.

Case in Point—Misaligned Metrics

A purchasing agent was sourcing axles for the bus company that he worked for. The mission of the bus company was Safety at all Costs." It focused its efforts on delivering the riders on time and "complete." The purchasing agent received a call from a new axle manufacturer located in Malaysia. This manufacturer offered a price significantly lower than all other manufacturers. After looking at the total cost of doing business with this new (untested) supplier, the buyer decided to take a chance and looked forward to a positive performance review. The axles arrived and were installed. Six months later a bus overturned causing many injuries. Upon inspection, the axle in the back of the base had cracked and split apart. The moral of the story is that often saving a dollar in the short term costs significantly more in the long run, especially if there is no connection to corporate objectives.

To overcome some of these challenges inherent in a performance measurement system, many organizations develop joint goals and performance measures that apply across the organization. For example, cost-savings goals must be identified usually in cooperation with other business units or functions. Purchasing is then responsible for achieving those savings goals.

New product launches are challenging for organizations because they are difficult to manage and cross many organizational units. One measure that has been developed relates to an on-time, new product launch.[4] This forces all the different functions to work together and commit resources to the launch. Each function has these measures included in its personal performance metric indicators. Some key measures used to measure an on-time, new product launch follows:

- Percentage of sales from products launched previous year
- Time to market (days)
- Products launched on budget
- Products launched on time
- Percent R&D cost for new products

Within the supply chain, many trade-offs in performance exist. Balancing these trade-offs requires the functional areas to pull together to focus on optimization of the supply chain instead of optimization of one particular function. This is easy to say, but difficult to implement. The trade-offs generally revolve around cost, service, and quality. However, because of increasing regulatory mandates and global warming, another trade-off is included on the list: carbon emissions (aka GHG) that are discussed throughout this book.[5]

Gaining a competitive advantage and meeting the ever-increasing demands of customers forced companies to implement strategies such as long-distance air-freighting, small batch sizes, just-in-time concepts, and energy-intensive production in countries with low environmental standards.[6] These strategies typically focused on the optimization of one company versus the optimization of supply chain efficiency. As shown in the section on "Strategic Cost Management," concerns for inventory levels and the cost of inventory were pushed back to the supplier level. There was little or no regard for the impact on the supply chain. The bullwhip effect has a similar influence. Poor planning and poor relationships with customers generate additional costs in the upstream supply chain.

The introduction of carbon into the service and trade-off equation means that companies have to rethink their geographic positioning of the supply base, modes of transportation, processes used in manufacturing, and others. The goal should ultimately be to optimize supply chain products, processes, information, and cash flow with the four trade-offs: cost, quality, service, and carbon. Traditional practices have to change with impending regulations and customer demands, which means that balancing of these trade-offs becomes more complex.

Optimization requires a holistic approach. The trade-offs should be evaluated in terms of their relationship to one another and impact on performance goals. Many options are available to reduce costs, carbon, and other performance measures simultaneously. Figure 6-1 shows this trade-off model with the different options.

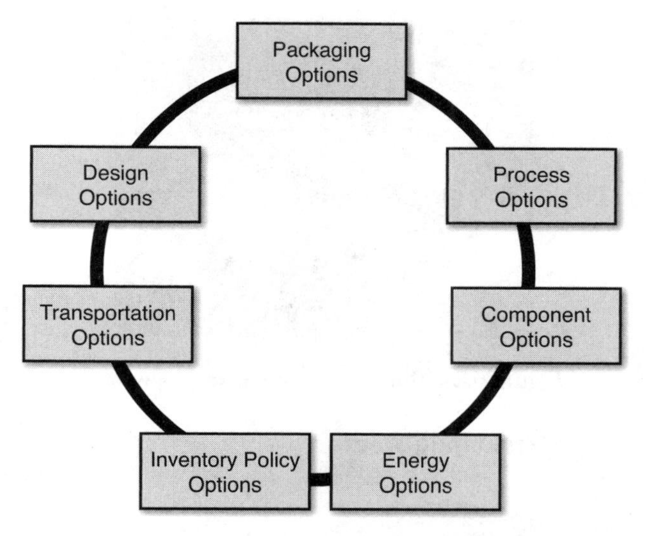

Figure 6-1 Trade-off options for improved performance[7]

Within each option, you can assess many items that influence the four trade-offs. The key again is to look at this holistically and try to understand the influence of each change on the supply chain and how these trade-offs are managed. Following are some ideas for each option to help achieve the right performance.

- **Design options**—You must consider many options during the design of products. The entire product development team needs to be aware of the impact of design decisions on supply chain optimization. Things such as selection of materials that go into the product, durability, and upgradeability should be considered. Packaging and transportation issues should be part of the initial design, not considered as an afterthought. Many "take-back" laws are being implemented by regulators, so consideration of ease of disassembly, recyclability, and disposability are important. Even the process of development could change to a more virtual mode to reduce extensive travel.
- **Packaging options**—Packaging waste is a worldwide problem and organizations are setting targets to eliminate, or at least reduce, the amount of waste generated by their products and processes. Things such as packaging size, packaging materials, reuse and recycling of packaging materials, substitution of packaging with creative transportation ideas, and documentation such as assembly instructions and usage manuals should be considered. Figure 6.2 is one of many illustrations

of the problem of packaging waste. Figure 6.3 gives a breakdown of the types of materials that are most common in the waste stream.[8]

Figure 6.2 Illustration of packaging waste[9]

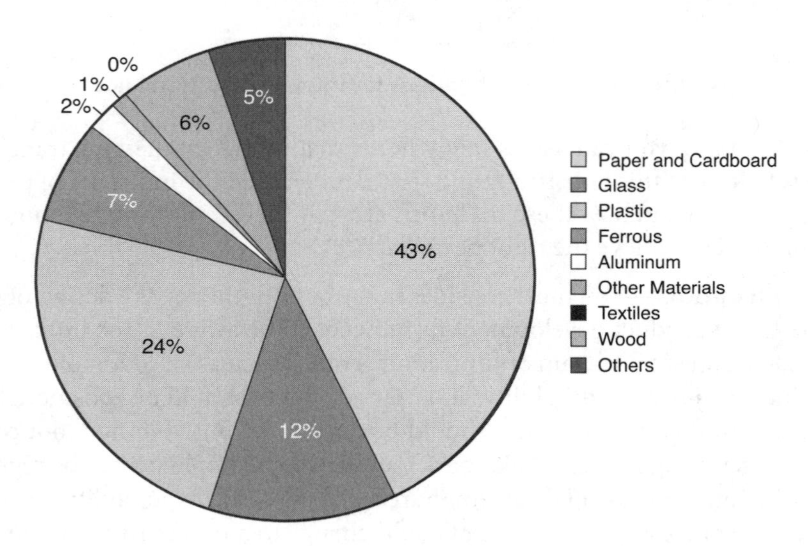

Figure 6.3 Types of packaging waste

- **Process options**—Process options look at processes used to plan, source, make, and deliver the products to the customer. This includes order fulfillment, manufacturing, transportation, quality control, and demand and supply planning. The processes are largely cross-functional, so decisions and metrics should have input from many different areas of the organization.

- **Component options**—The components that go into products are largely managed by the purchasing area but need to have inputs from many other parts of the organization. For example, supplier location is a critical decision, and often it is easy to find the low price but at what cost to the organization? Supplier location also influences transportation, inventory, and availability. Purchasing could consider substitute components made from different material or using a different process. Bringing the product or component back in-house in some respects helps to better manage costs. Supply base optimization and rationalization (discussed earlier) is another way to influence the appropriate performance in cost, quality, service, and carbon.
- **Energy options**[10]—There are many alternative energy sources that are fossil fuel-based (oil and natural gas); renewable energy-based (ethanol, solar, and wind); and other (nuclear and geothermal). Companies need to seek out these alternative uses and try to reduce the fossil fuel-based options. Companies should also seek out suppliers working to reduce fossil fuel-based options and also reduce energy usage in general. Over a 10-year period, BP is investing $8 billion in the use of alternative fuel sources.[11] Siemans is an award winning user and developer of renewable energy sources with a focus on wind, distributed and hybrid energy, hydro, solar, and biomass power sources.[12]
- **Inventory policy options**—Logisticians, marketers, and purchasing professionals have long been at odds regarding inventory policies. Consideration of policies on safety stock, lot sizes, and planning frequency are important. Planning horizons and change policies have a significant impact on inventory levels. Replenishment programs from both the supplier and to the customer should also be part of the trade-off assessment.
- **Transportation options**—Transportation has a large influence on cost, quality, service, and carbon. Figure 6-4 shows that greenhouse gas emissions account for 28 percent of the total U.S. GHG emissions, making it the second largest emitter behind the electricity sector.[13] Taking this to a global scale magnifies the problem. Location of distribution centers and the modes used in getting the products that are needed make a difference. Shipping frequency, partial shipments, load consolidation and routing also are a concern.

Obviously, you must consider many trade-offs when trying to balance cost, quality, service, and carbon metrics. The company must take a holistic, supply chain perspective and think about reward structures and the type of behavior that these measures initiate. You need to establish the appropriate goals, and those goals need to be aligned both horizontally and vertically. The challenge is finding a mix of performance measures that incorporate cross-functional objectives.

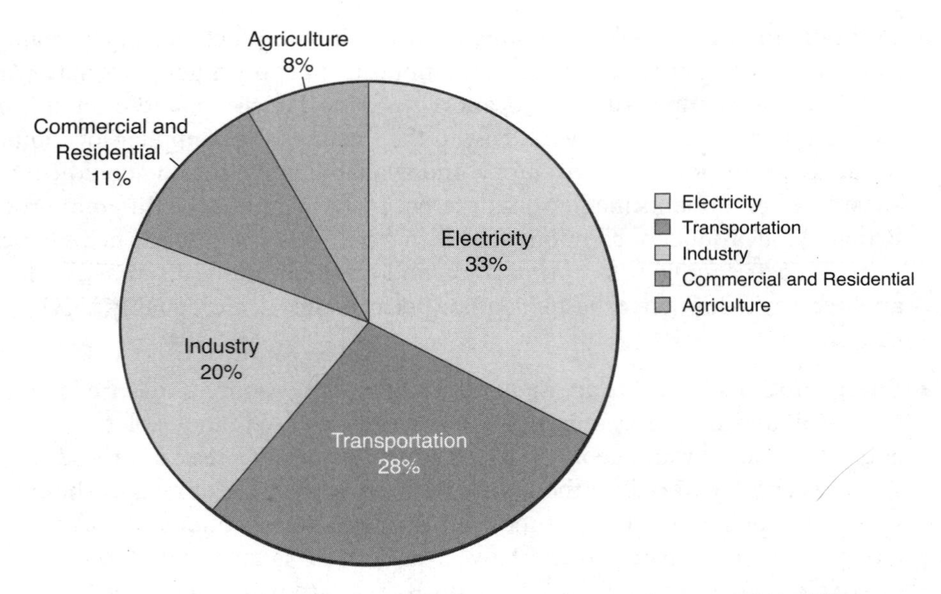

Figure 6-4 Total U.S. greenhouse gas emissions by sector (2011)[14]

Types of Measures

Many different types of measures have been applied to the supply management area. These measures are divided into three primary categories: efficiency metrics, effectiveness metrics, and integrative metrics.[15] All these metrics are meant to focus purchasing on meeting the goals of the organization, goals of the business units and other functions, and goals of the department. The following sections discuss a number of broad categories of measures, followed by some specific integrative measures. The difficulty of performance measurement is that it is not a one-size-fits-all process. Measures that encompass the tactical aspects of purchasing's responsibility are relatively straightforward. However, measures that include purchasing's involvement with other functions or teams are difficult to implement and manage.

Efficiency Measures

Measure of efficiency are about "doing the thing right."[16] Administrative costs are the basis for measuring purchasing efficiency. These measures relate to how well the purchasing department performs in the activities it is expected to perform against the budget that is in place for the department.[17] If the purchasing costs are within the budget, the efficiency of the purchasing department can exceed expectations. These measures typically focus on price and department operating efficiency. The idea is to compare the results of these calculations with similar figures for previous time periods. This can help to evaluate the efficiency of the supply function. Performance measures here include material price reductions, operating costs, and order processing time. Purchase price variance was discussed in earlier chapters. This is a typical purchasing efficiency measure.

Effectiveness Measures

Effectiveness measures are also about doing the right thing.[18] Effectiveness measures are a bit more challenging because of market fluctuation in pricing and seasonal variability impacting inventory. Effectiveness measures attempt to measure how well something is done. These measures can include evaluating direct and indirect contributions to final customer satisfaction, profit, revenue enhancement, or asset management. These measures might also include the quality of supplier relations or levels of internal customer satisfaction.

Purchasing Functionality

Purchasing performance can be measured against the functional requirements of the purchasing function. The primary function of the department is to provide the correct item at the required time at the lowest possible cost.[19] This is a metric that is also challenging to implement because of other factors, such as supplier stability, material quality issues, and supplier discounts that are not considered.

Examples of Integrative Measures of Purchasing Performance

Examples of integrative measures of purchasing performance include

- **Return on net assets (RONA)**—Return on net assets considers fixed assets (real estate and machinery) and also net working capital (current assets and current liabilities).[20] The higher the RONA, the better the company performance. Most of the primary functions of an organization influence the elements considered in calculating RONA. To improve the return on net assets, an organization needs to introduce new products regularly with the features and the price points that the customer demands. The costs of existing products also need to be reduced while maintaining margins. The third piece of this is managing net assets. To achieve an improved RONA purchasing, the business units must work together to establish cost-saving targets.
- **Total cost of ownership**—Total cost of ownership has been discussed in many of the earlier chapters. Some organizations track and include in the total costs elements from the products point of origin for goods all the way to the point of where that materials reaches the back door of the retail locations.[21] Also included is the reverse flow of the products in the case of mandatory take back or disposal. However, using total cost as a performance measure involves many resources and

inputs, so careful consideration of the material importance of the element to the total cost should be taken.

Tracking a cost that is not a significant percentage of the total is good, but more important, the big cost elements need to be well understood. Also comparing costs from one point in time to another needs to include the same or similar cost elements. The total cost of ownership is an integrative measure that should be used when determining the source of supply and also for measuring and managing performance.

- **Technology and innovation**—This might include measures of new technology from the supplier and supplier suggestions for product and process improvement. Annual supplier spending on research and development and the number of patents applied for and issued would be something else to consider in this category. This also may include improved efficiency of the purchasing department: EDI, e-procurement systems, VMI, and other technologically innovative processes.
- **Availability measures**—These include not only the timing and the quantity of orders, but also the availability of inventory. When inventory is included as part of the measurement of many functional areas, there is less likely an opportunity for people to work in their own best interests.
- **Quality measures**—Include quantitative quality measures, user surveys, and supplier quality certification programs. This is a good measure for many areas as well. For example, purchasing would be about finding the right supplier; engineering would be about designing a product to the appropriate level of quality.
- **Customer satisfaction measures**—Purchasing's internal customers are important. Some of these customer service measures include contract execution time, customer satisfaction, contracting process satisfaction, purchasing and client communications, responsiveness and cooperation, purchasing knowledge and capabilities, overall purchasing team performance, and internal and external customer satisfaction survey results.

Following is a compilation of many of the metrics used to assess purchasing performance. As usual, this is not comprehensive but includes many of the most frequently occurring metrics.

- Market price trends compared to prices paid suppliers
- Cost and price variances
  - Target price measures, cost reduction measures, cost avoidance
- Transportation costs
- Lead time and delivery expected compared to actual

    - On-time delivery (supplier delivery performance)
    - Complete delivery
- Inventory measures
    - Purchase inventory dollars
    - Inventory investment changes, inventory turnover
    - Holding and obsolescence cost for inventory
    - Planned versus actual inventory
    - Fill rate, inventory accuracy
- Quality
    - Defective parts per million
    - Percentage and number of defects
    - Customer quality incidents
    - Dollars recovered from suppliers because of poor quality
    - Impact and cost of defect
    - Quality system improvements
- Change orders processed
- Purchase orders issued
- Employee workload and productivity
- Average dollar cost of the purchase orders written
- Operating costs as a percentage of total dollar volume of purchases
- Operating costs as a percentage of total dollar volume of sales
- Comparison of actual department operating costs to budget
- Cash discounts earned and cash discounts lost
- Internal customer satisfaction
- Key supplier problems that could affect supply

There are many common problems with the more traditional measures and measurement systems. One significant concern is that the annual assignment of goals tends to inhibit continuous improvement processes.[22] Many of the metrics are also easy to develop and more easily managed. However, they do not lead to actual process improvement, for example, measuring the number of purchase orders issued. This type of measure would reward behavior such as smaller orders and more frequent deliveries. The other problem is that these types of measures tend to produce conflicting results because they are inconsistent across the organization. Many of these existing measures are also tactical rather than strategic.

Measurement Systems and Frameworks

There is much research available that tries to understand the ideal measurement system. Many frameworks such as the balanced scorecard have been presented and adapted over time. Some basic guidelines for the ideal measurement system include

- Support the business's operating goals, objectives, critical success factors. and programs.
- Provide as simple a set of measures as possible.
- Reveal how effectively customers' needs and expectations are satisfied.
- Allow members of each organizational component to understand how their decisions and activities affect the entire business.
- Support organizational learning and continuous improvement.
- Provide congruency of measures across organizational levels.[23]

Measures should be action-oriented and timely. Measures should also be adapted for the appropriate organizational level. At the lower levels of the organization, measures should lead to immediate, operational solutions. At the middle-management level, measures should invoke changes in operational procedures or focus. For top management, measures should indicate changes in the choice of strategies to meet goals.

Many organizations approach the measurement process looking for measures that are both comprehensive and meaningful. One way to do this is to develop a scorecard that measures the areas that are important to the company, the SBUs and other functions, and purchasing. These categories are both quantitative and qualitative and have many individual metrics included within the category. Some of the primary categories that are included on the scorecard are shown in Table 6-2.

Table 6-2: Scorecard Categories to Measure Purchasing Performance

Primary Scorecard Category	Description
Shareholder value enhancement	Financially oriented metrics that focus on both the top and the bottom line of the organization.
Procurement as a business partner	Working with internal customers to meet their needs.
Procurement organization productivity	Efficiency metrics, delivery, inventory, and quality. This includes execution of critical strategies and continuous improvement.
Talent management	People-related measures: Are employees being recruited and developed? Are employees satisfied?

Other measures integrate the supplier into the success of the organization and the supply chain. One such measure is the Relative Value Index of the supplier, which includes the amount paid to the supplier, the cost of quality, the cost of flexibility, and the cost of risk. This compares the amount paid to the supplier and the supplier's impact on the business.

Organizations have also developed policies to identify true-value add by procurement. In most cases, these relate to cost-savings. However, cost-savings have to be perceived as valid and believable by others in the organization. In other words, procurement has to take some action that results in reduced spend or cost avoidance.

Conclusion and Chapter Wrap-Up

Many organizations have shifted their focus from labor hours and purchase price variance to more integrative measures such as

- Cost reduction
- Total cost
- Lead times
- Quality
- Inventory
- Order to delivery (OTD) time
- Responsiveness to customers

The measures that executive supply management professionals are concerned with fall primarily into eight different categories.

- Supply's contribution to innovation
- Profit and loss (P&L) impact of supply
- Risk mitigation
- Revenue contribution
- Growth
- Accuracy of predictive indicators
- "Soft" areas such as professionalism and effectiveness of training
- Supply management competencies and skills

Performance measures have to meet the needs of executives, business units and other functions, and also the individual. Development of the metrics is critically important; however, managing the metrics that matter is even more critical. Figure 6-5 shows a six-step process for developing relevant purchasing performance measures.[24]

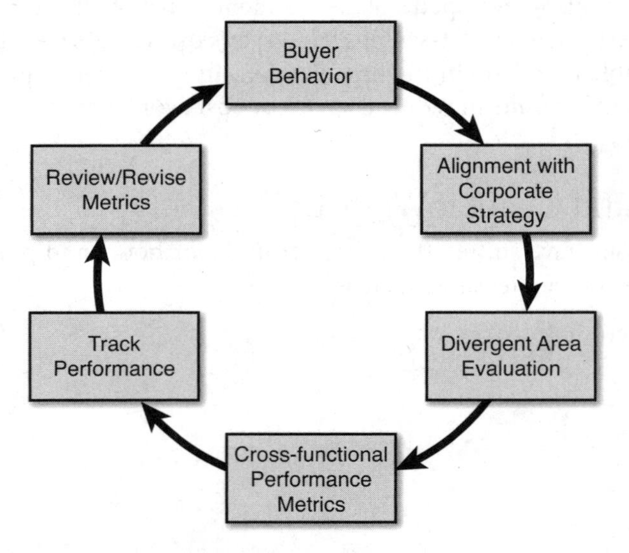

Figure 6.5 Purchasing performance measures

The purpose of this chapter is to gain an understanding of the internal and external performance assessment of supply chain and supply management operations. Consideration of the role of measurement, different types of measures, and the various trade-offs associated with performance metrics are discussed. The primary takeaway is that integrative measures that drive the appropriate behavior are the most effective. You must develop the right metrics and manage those metrics in a way that makes sense. Measures that are both qualitative and quantitative need to be established.

In summary, there are a number of skills and capabilities related to performance measurement that supply managers should consider developing as they attempt to move their organizations to increasingly better financial and market positions. These include

- Understand the role of measurement to establish quantitative purchasing goals and key performance indicators.
- Develop integrative performance metrics that measure purchasing performance on other areas of the organization.
- Understand the different categories of supply management performance measurement and key performance indicators.
- Apply scorecards to the supply management organization.

Key Terms

- Purchasing
- Supply management
- Performance measures
- On-time new product launch
- Total cost focus
- Inventory control function
- Efficiency measures
- Effective measures
- Integrative measures
- Inventory measures
- Return On Net Assets
- Total Cost of Ownership
- New product development
- Technology and innovation
- Availability measures
- Quality measures

- Customer satisfaction measures
- Shareholder value enhancement
- Procurement as a business partner
- Procurement organization productivity
- Talent managers
- Relative value index

References

1. ESResearch (2008). Enabling Analysis. Retrieved September 19, 2013, from http://www.esresearch.com/e/downloads/EnablingAnalysis_Synygy_Summer08.pdf.
2. McKenzie, G. (2013). "If you can't measure it, you can't improve it." Retrieved on September 19, 2013, from http://guavabox.com/if-you-cant-measure-it-you-cant-improve-it/.
3. Monczka, et al. (2011), ibid.
4. AFMA (2004). "Key Performance Indicators." Retrieved September 11, 2013, from http://www.ahfa.us/uploads/documents/Performance.pdf.
5. Butner, K., Gueder, D., Hittner, J. (2008). "IBM, Mastering carbon management. Balancing tradeoffs to optimize supply chain efficiencies." Retrieved September 11, 2013, from http://www-05.ibm.com/de/automotive/downloads/mastering-carbon-management.pdf.
6. Ibid.
7. IBM Research and Institute for Business Value. Retrieved September 24, 2013, from http://www-03.ibm.com/press/us/en/pressrelease/24296.wss.
8. Green Print Survival. "Waste Management—Top Down Approach." Retrieved September 11, 2013, from https://www.google.com/search?q=packaging+waste&tbm=isch&tbo=u&source=univ&sa=X&ei=T24wUqPPGoSH2gW0voGoAQ&ved=0CGIQsAQ&biw=1044&bih=732#facrc=_&imgdii=_&imgrc=_5E-gg8IUZKeiM%3A%3BH5JzhFJQ8rLN9M%3Bhttp%253A%252F%252Fgreenprintsurvival.files.wordpress.com%252F2008%252F09%252Fpackaging-waste1.jpg%3Bhttp%253A%252F%252Fgreenprintsurvival.wordpress.com%252F2008%252F09%252F15%252Fwaste-management-top-down-approach%252F%3B378%3B284.

9. EIONet. "What is waste?" Retrieve September 11, 2013, from https://www.google.com/search?q=packaging+waste&tbm=isch&tbo=u&source=univ&sa=X&ei=T24wUqPPGoSH2gW0voGoAQ&ved=0CGIQsAQ&biw=1044&bih=732#facrc=_&imgdii=_&imgrc=mR49VszW8ZWYCM%3A%3BiOCBFcYC8zA9xM%3Bhttp%253A%252F%252Fscp.eionet.europa.eu%252Fthemes%252Fwaste%252Ffigures%252Fphoto_packaging%3Bhttp%253A%252F%252Fscp.eionet.europa.eu%252Fthemes%252Fwaste%3B385%3B303.

10. Alternative Energy. Alternative Energy News. Retrieved on September 11, 2013, from http://www.alternative-energy-news.info/.

11. BP. Alternative Energy. Retrieved September 11, 2013, from http://www.bp.com/modularhome.do?categoryId=7040&contentId=7051376.

12. Siemans. Siemans Renewable Energy. Retrieved September 11, 2013, from http://www.energy.siemens.com/us/en/renewable-energy/?stc=usccc025800.

13. EPA (2011). Sources of Green House Gas Emissions. Retrieved September 11, 2013, from http://www.epa.gov/climatechange/ghgemissions/sources/transportation.html.

14. EPA, ibid.

15. Monzcka, et al (2011).

16. Chaffey, D. (2011). "What is the difference between efficiency and effectiveness measures?" Retrieved on September 11, 2013, from http://www.smartinsights.com/goal-setting-evaluation/goals-kpis/definition-efficiency-and-effectiveness/.

17. Murray, M. (2013). "Measuring Purchasing Performance." About.com—Logistics/Supply Chain. Retrieved September 4, 2013, from http://logistics.about.com/od/strategicsupplychain/a/measure_purchasing.htm.

18. Chaffey, D. (2011), ibid.

19. Murray (2013), ibid.

20. Investopedia. (2013). Return on Net Assets (RONA). Retrieved September 11, 2013, from http://www.investopedia.com/terms/r/rona.asp.

21. Business Dictionary. Total Cost of Ownership. Retrieved September 11, 2013, from http://www.businessdictionary.com/definition/total-cost-of-ownership-TCO.html.

22. Raedels, A., Buddress, L. (1998). "What Is Purchasing Success and How Do We Know If We Did It? ISM 83rd Annual International Conference Proceedings. Retrieved September 4, 2013, from http://www.ism.ws/pubs/proceedings/confproceedingsdetail.cfm?ItemNumber=10832.
23. Raedels and Buddress (1998), ibid.
24. Fearon, H.E., Bales, W.A. (1997). "Measures of Purchasing Effectiveness," Tempe, AZ: Center for Advanced Purchasing Studies.

7

CONCLUSIONS AND TRENDS IN SUPPLY MANAGEMENT

As seen throughout the text, there are dramatic changes occurring in the area of supply and supply chain management. The purpose behind this book was to provide an overview of the purchasing function, its role both internally and externally to the organization. In addition, some of the basic tools and technologies used to facilitate an uninterrupted flow of products and services in the most efficient and effective manner were also introduced. One key role of a supply manager is to help an organization mitigate risk and ensure that products and services are purchased at the right price, place, time, and condition. This final chapter discusses some of the key themes occurring throughout the book, looks at future trends in supply management, and focuses on essential skills needed to be successful.

Learning Outcomes

After completing this chapter, you should be able to:

- Identify some key trends in the supply management area.
- Think about the types of skills required to be a successful supply manager.

Conclusion

There were many ideas introduced throughout each chapter that purchasers need to be cognizant of. These ideas can help a supply manager be successful and also to be a better cross-functional team member. The contribution that a supply manager can make to organizational success is significant. This section discusses some of the key concepts in each chapter and then directs you to those integrative concepts that crossed multiple chapters.

Chapter 1, "The Essential Concepts of Purchasing and Supply Management"

Chapter 1 introduces the essential concepts of purchasing and supply management and shows how this function has moved from a transactional-oriented function to one that is a strategic contributor. This shift in orientation enables the purchasing function to influence both the top and the bottom line of the organization.

Globalization has been one of the key drivers for companies to improve their internal processes such as supply management. To be successful it is no longer an issue of "lowest price," but instead it is about delivering total customer value. Technology has also played a key role in linking together members of the supply chain and helping to facilitate getting the product to customers at the right time, place, cost, and quality.

Another key idea in Chapter 1 is that by working with its suppliers, purchasing adds value to the organization by improving quality, increasing the adoption of innovation, and decreasing time to market. Purchasing also excels at cost reduction opportunities. Furthermore, organizations are now strategically taking a more holistic approach to supply base management and cost management. Purchasers now look total and life-cycle costs of doing business with a supplier versus trying to achieve only cost or price reductions.

Purchasing also has many interfaces within the different organizational units. Chapter 1 also discusses some of these key interfaces and the different activities that take place when the functions are required to interface.

The chapter also outlines the basic steps in the purchasing process. Kraljic's matrix was first introduced as a way to help move the purchasing process forward in an effective and efficient way. Both strategic and tactical roles of the purchaser are mentioned, and these set the stage for the later chapters.

Chapter 2, "Key Elements and Processes in Managing Supply Operations and How They Interact"

Chapter 2 introduces the concept of commodity strategy and the supply management processes. An important aspect of purchasing is to execute purchasing strategies in a way that protects the organization from operational, financial, and reputational risk. Many of the high-impact supply chain operational risks are discussed.

An important first step in mitigating this risk and moving forward with the sourcing process is to understand what you have, what you need, and what is happening on the external market. First, the purchaser needs to perform an assessment of spend and demand. This introduces the procure to pay idea that was simplified by the introduction of e-sourcing tools.

Porter's Five Forces model is also introduced in this chapter as a way to assess market conditions[1] and as a way to organize the data collected in the spend and market analysis. Kraljic's matrix is reintroduced and discuss more in-depth as a way to perform market analysis.

Supplier analysis looks at both current and potential suppliers and can be facilitated through the use of a scorecard or ranking system. The most common system is a weighted scorecard to ensure that the strategic needs of the buying organizations are met.

The chapter closed by discussing the need for better contract management. Supply managers must have the capability to effectively prepare and negotiate contracts that are in the best interest of the buying organization. However, these contracts must be managed in a way that also allows for supplier development and growth. There is a need for better contract management in general, and supply managers are positioned in a way to better develop and manage the contracts.

Chapter 3, "Principles and Strategies for Establishing Efficient, Effective, and Sustainable Supply Management Operations"

Chapter 3 introduces supply management principles and strategies by again incorporating Kraljic's portfolio matric. This chapter finds the segmentation and classification process useful in helping to understand how to manage the suppliers of particular commodities and services. The importance of establishing a highly functioning cross-functional team was introduced; supply managers need the relational skills to work with on these teams to work with their internal customers and coordinate with suppliers.

The key sourcing strategies such as supply base optimization, insourcing and outsourcing, offshoring, reshoring and nearshoring, which are all popular in the practitioner press, are addressed. One of the main concepts and key strategies introduced in this chapter is the notion of managing supply chain costs. Ultimately, managing costs across a complex and constantly adapting supply chain is challenging; therefore, some strategic cost management tools are introduced. Other supply chain analytics that help guide decision making are also mentioned. Supply managers must use these tools in the appropriate way so that the costs across the supply chain are well managed.

Another overarching theme is that of sustainability and purchasing's involvement in sustainability efforts. Sustainability efforts across the supply chain are challenging because of different regulations, customs, and customer demands. The primary trade-offs of cost, quality, and service are introduced and a fourth item—carbon—is added. Sustainability—environmental, economic, social—will continue to grow in importance and the involvement of supply management is critical.

Chapter 4, "The Critical Role of Technology in Managing Supply Operations and Product Flows"

The focus in this chapter is on technology that helps to facilitate the job of purchasers. The transition from internally focused systems and software (MRP, DRP, and ERP) set the foundation for later externally focused systems such as e-sourcing. The capabilities of supplier relationship management e-sourcing suites are mentioned. The most interesting part of this chapter is the many changing and challenging trends that are identified with the discussion on how these trends will also impact the role of supply managers. A number of e-sourcing tools are discussed and some future technology trends that impact supply managers are mentioned.

Chapter 5, "Define the Requirements and Challenges of Sourcing on a Global Basis"

Chapter 5 moves the sourcing process into the global arena and compares and contrasts the differences between sourcing globally and sourcing domestically. The advantages and disadvantages of global sourcing and how these relate to the job of supply manager are discussed. Data is presented that shows the trends in sourcing from different regions of the world.

Many additional challenges associated with global sourcing include cultural, language, communication, and social customs. Purchasers must be culturally astute before doing business in different parts of the world. This chapter shows the importance of the total cost of ownership, in-depth supplier analysis, assessment, and country location analysis. The overarching goal in a global context is for purchasers to better mitigate risk and ensure compliance with carefully developed contracts.

Chapter 6, "Assessing the Internal and External Performance of Supply Management Operations"

This chapter look at the role of metrics and performance measures in supply management operations. You gain an understanding of the internal and external performance assessment of supply chain and supply management operations. The interesting perspective of this chapter is to understand the right types of metrics, how to develop metrics that drive the right behavior, and what metrics are effective for teams and also intra-organizationally. The idea is not to measure what you can but instead measure what is needed. Metrics should be developed that take a more balanced approach and drive the appropriate behavior. This chapter discusses the many trade-offs associated with performance and how to develop metrics that incorporate the needs of the organization.

Summary of Key Concepts That Cross Multiple Chapters and Supply Management processes

There are some overarching key ideas that crossed many, if not all, the chapters. These key concepts are important for current and future supply managers to understand:

- **Risk management**—A comprehensive risk management strategy helps to identify, assess, and prioritize risks of different kinds. Purchasing managers must create a plan to minimize or eliminate the impact of negative events.[2] There is risk associated with all aspects of the sourcing process. Purchasers need to assess financial risk, product risk, reputational risk, and other risk to ensure that their organizations maintains a competitive advantage.
- **Total cost and life-cycle costs**—Total cost modeling helps to determine the total cost of doing business with a specific supplier especially in situations in which geographic decision making is involved. Total cost modeling enables you to look at different sensitivity analyses to understand the what-if consequences. Furthermore, it is also used as an integrative cross-organizational metric and a measure of purchasing performance.
- **Category/portfolio management**—Each chapter includes a discussion on how the segmentation and classification of categories of services and products can help a purchaser manage the supplier in an appropriate fashion. This matrix helps a purchaser to classify commodities and services into categories based on importance of the purchase to the organization and the complexity and risk associated with the supply base. This process of classification helps purchasers to identify the appropriate suppliers, determine the type of supplier relationship, and better manage the commodity using commodity strategies, tactics, and actions.
- **Integration**—There are many points of integration for purchasing. This is partially because purchasing is a cross-functionally oriented process and because of its boundary spanning capabilities. There are many points of integration within teams, internally (other functions, roles, and business units); externally (suppliers and customers); technologically (add-ons, legacy, supplier, and customer); and in performance measures.
- **Scorecarding and performance measurement**—Developing the appropriate tools for measuring supplier performance, identifying suppliers, and measuring supply management is critical. Scorecarding is one way to develop the criteria and weight the criteria depending on importance and fit with organizational strategy.

Future Trends

Purchasing has made great strides in moving from a tactical buyer to a strategic contributor in an organization. Figure 7-1 has a number of the activities that help make this shift possible.

Figure 7.1 Moving from tactical buyer to strategic contributor

This shift from tactical to strategic has helped to improve the competitive advantage of an organization, has increased the visibility of the supply management organization, and has allowed supply managers to be involved in introduced and implementing strategies that could potentially change the face of business. Both researchers and practitioners focus their efforts on these potentially game-changing trends. The following list shows some key trends, but is not all-encompassing. However, it gives you a good idea of the areas that "keep managers awake at night."

- Supply managers must develop risk management and mitigation strategies and tools that protect the company from financial, operational, and reputational risk. This requires an increased presence in sustainability initiatives: environmental, social, and economic.

- Supply managers must develop and improve value-focused sourcing and value chain analytics.
- Using the new technologies, including e-sourcing, social networks, and cloud computing, supply managers must increase information sharing and transparency internally and with suppliers. This includes enhanced supply chain integration and collaboration with suppliers to incorporate innovation efforts across the value chain.
- The manufacturing and supplier location decision requires supply managers to assess insourcing, reshoring, and offshoring efforts with a specific focus on risk management.
- Supply managers must involve suppliers in cost-management activities that include the product, process, and supply chain initiatives.
- Improve cross-functional integration with other functions, processes, and customers including customer-facing activities.
- Supply managers must recruit and manage talent effectively and efficiently.

Essential Skills

Talent management is a key concern of executives in the supply area. Purchaser's need to have a good balance of both relational skills and analytical skills. Additional key skill-sets necessary in this area include market and supply process expertise. Furthermore, purchasers must understand the competitive market structure and apply the appropriate pricing and costing models to prepare for analysis. In addition, purchasers must also have knowledge of future forces that impact their commodities and markets, such as emerging supply markets, sustainability, potential new competitors, and pending merger and acquisition.

Companies must look for talented, flexible, globally oriented personnel who can immediately contribute to an organization's competitive advantage. These people must have cross-functional and teaming skills, cross-cultural perspectives, and innovative and leadership skills.

This book has addressed a single-learning block and has helped increase your awareness of purchasing's strategic role in the organization. No matter what position you are in, the purchasing organization is cross-functional and touches many areas of the business. There are still many ways that purchasing can contribute and improve supply operations.

References

1. Porter's Five Forces Model is a commonly used managerial tool first introduced in Michael Porter's (1979), "How Competitive Forces Shape Strategy," *Harvard Business Review*, March/April.
2. Anonymous. "What Is Risk Management?" Retrieved September 4, 2013, from http://www.whatisriskmanagement.net/.

Index

A

B

C

D

E

F

G

H

I–J

K

L

M

N

O

P

Q

R

S

T–U

V

W–Z